SOTHIS PRESS

Sothis Press aims to completely reimagine the public domain book with an emphasis on universal access to captivatingly designed print editions. We believe that technology will enable every single book ever published to be perpetually accessible in beautiful print form. Sothis is beauty in ubiquity.

www.sothis.press
1 Waterhouse Square, London, EC1N 2ST

Modern Steam Traps (English and American)

Their Construction and Working

Gordon Stewart

SOTHIS PRESS

ISBN 978-1-0368-0215-8

This edition is published by
Sothis Press © 2023. The company
is registered in England and Wales.
The Company Number is 14900389.

MODERN STEAM TRAPS

(ENGLISH AND AMERICAN):

THEIR CONSTRUCTION AND WORKING.

BY

GORDON STEWART.

PRICE THREE SHILLINGS NET.

THE TECHNICAL PUBLISHING CO. LIMITED,
55 AND 56, CHANCERY LANE, LONDON, W.C.;
D. VAN NOSTRAND CO., 23, Murray Street, and 27, Warren Street, New York;
E. W. COLE, Sydney and Melbourne, Australia;
GEO. ROBERTSON AND CO. PROPRIETARY LIMITED, Melbourne, Sydney, Adelaide and Brisbane; and all Booksellers.
1907

PREFACE.

This work first appeared as a serial in the pages of *The Practical Engineer*, and the interest created encouraged the idea of republishing same in book form.

In placing the book before the public it is perhaps advisable to add a few words by way of explanation. There is an undoubted want of knowledge among users of Steam Traps concerning the points which are most vital to the proper and efficient working of such apparatus; and while this book does not profess to deal exhaustively with the subject, it is hoped that the reader will allow a small amount of indulgence, and remember that this is the first work published dealing exclusively with Steam Traps.

This opportunity is taken of thanking the numerous makers for the information, photos., etc., which they have been kind enough to loan.

GORDON STEWART.

London, 1907.

CONTENTS.

LIST OF ILLUSTRATIONS.

MODERN STEAM TRAPS.

CHAPTER I.

INTRODUCTION.

IT is common knowledge amongst steam engineers that unless proper means are taken to reduce to a minimum the possible degree of wetness of the steam used by a reciprocating engine the efficiency of the plant is considerably reduced, and in some cases the engine is placed in imminent danger of a serious breakdown; as, for instance, the cylinder cover of a high-speed engine may be blown off, due to an accumulation of water in the cylinder; for clearance in a high-speed engine of modern construction is always very small.

First of all let us consider why it is that water can find its way into the pipes and cylinders of an engine. Steam is spoken of as being either dry, superheated, or wet. Dry steam is the result of raising water to boiling point, and by a further expenditure of heat (latent heat) converting it entirely into steam at that temperature. Superheated steam is steam which is heated above the temperature of that due to the pressure at which it was formed, whereas wet steam contains minute particles of water which, instead of being converted into steam, have passed out of the boiler with it as it rushes to the engine; this action is spoken of as entrainment. There is also another reason for the presence of water, that due to condensation, which always takes place in every steam pipe system.

The possibilities of wet steam reaching an engine may be reduced in three ways: (I) By constructing the boiler so as to decrease the possibility of the water in the form of spray becoming entrained with the steam. (2) By superheating the steam as it leaves the boiler. (3) By automatically separating the entrained and condensation water from the steam immediately before it enters the cylinders.

If the steam space in a boiler is relatively small, and especially if the surface area is small, considerable disturbance is caused by the steam in liberating itself from the water, which disturbance is much increased if the boiler is forced or if the water is of a scummy nature; so that the spray which is formed at the steam liberating surface becomes entrained with the steam and passes away with it to the engine. It is possible to reduce this entrainment by considerably increasing the steam space and the steam liberating surface, but a practical limit is very soon reached in this direction; so that at its best only a compromise between the two, compatible with considerations of both cost and design, may be arrived at. However, if the wet steam as it passes from the boiler is led direct to a superheater, the entrained water will be converted into steam, but at the expense of the efficiency of the superheater; moreover, wet steam very rapidly deteriorates a superheater. Some of the earlier economies, said to be due to superheating when low pressures were in vogue, may in the light of more accurate knowledge be traced more directly to the drying of the steam than to the actual superheating.

After every possible means has been adopted to prevent entrainment, our attention should then be directed to means by which condensation in the pipes leading to the engine may be reduced to a minimum. This may be done to some extent by protecting the pipe with an asbestos or some other suitable pipe covering. But do what we may condensation and entrainment are always present to a greater or less extent. So that we have to resort to a method of automatically separating the entrained and condensation water from the steam before allowing it to pass to the cylinder; this is done by passing the wet steam through a separator, steam drier, or water snatcher, as it is variously called.

It is not proposed to deal at all fully with the subject of

separators, beyond including a short description, by way of explanation, of one which depends for its working upon the principle of centrifugal force, the only principle of any marked success—the "Stratton" separator in fig. 1. Steam entering at the inlet on the right-hand side meets with a deflecting plate, which directs it round the central

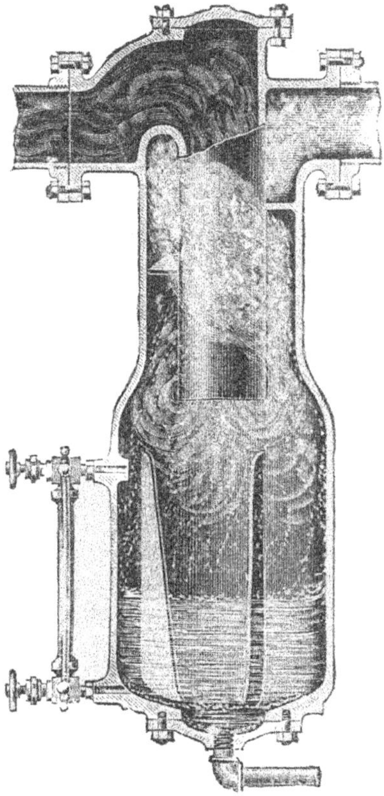

FIG. 1.—The Stratton Separator.

tube, imparting to it a rotative motion of great velocity; so that the heavier particles entrained with the steam, such as water, grit, etc., fly by centrifugal force to the walls of the outer cylinder, to which they adhere and flow down into the receptacle below. When the water thus accumulated is drained away by means of an ordinary steam trap, the dried steam passes up through the central tube to the outlet. The projecting vanes at the sides of the receptacle are for the purpose of slowing down the rapidly moving particles of water and grit, so as to eliminate the possibility of them being again swept up by the outgoing dried steam.

Few fittings of general utility in engineering workshops and factories have received such a vast amount of ingenuity and engineering skill of a high order expended upon their design, etc., as have modern steam traps. They have made their influence felt in almost every branch of manufacture, and now being an integral part of every installation they may be seen almost everywhere in daily use draining steam-pipe ranges—this being perhaps the most extended use to which they have been put—draining steam heaters, steam-jacketed kettles, steam presses, engine cylinders, steam-jacketed cylinders, paper-machine cylinders, air and gas receivers, vulcanisers, railway carriage heating appliances, etc.

It will be advisable to first fully discuss the features which go to form a good trap, and afterwards, by references to and descriptions of typical examples, to explain how these good points are obtained, and how the bad points are guarded against, leaving the reader to decide for himself which type of trap is most suitable in design and principle for the purpose he may have in view.

In these days of rush and competition, engineering design and practice resolves itself, like everything else, into nothing more nor less than the oft-repeated "survival of the fittest," and it is here that we see some of the old designs fast falling out, making way for new recruits, to stand or fall, as their predecessors did. For reasons that we need not enter into here, the very early designs of steam traps did not by any means prove reliable or efficient, due perhaps chiefly to undue strain and wear, and the use of unsuitable materials; many of them also had too many moving parts, with their attendant complications. The modern trap has, however, arrived at a very high degree of design and efficiency; so much so, that the difficulty that the engineer of to-day has to contend against, when in the market for traps, is to select the one which is most likely to suit his class of work, and to make his choice independent of the makers, for nowadays everybodies' products are "the best."

In selecting a steam trap for removing condensed steam from pipe ranges or similar work, the following points should be carefully watched for: By far the most important feature in determining the life and efficiency of any trap is whether it is or is not addicted to the vile habit of dribbling. There is no greater evil that a trap can possess, but with the amount of engineering skill and design brought to bear upon this point there is scarcely a trap on the market worthy of the suffix "Modern" which, when properly understood and once adjusted, can have this objectionable feature laid at its door. Prompt closing of the valve is the secret of the whole matter, for should a trap, after having discharged, remain open for even a very little while, steam will blow through, and, as is well known, the action of a steam blast upon the valve and its seating is even more detrimental than that of a compressed-air blast, cutting as it does both the valve and its seating every time discharge takes place. Should the trap be working under these conditions for any appreciable time, the valve and its seating will be ultimately ruined, and become at its best an exceedingly bad fit, so that the trap would always dribble and discharge live steam. A good trap should discharge in gusts, for gusts are audible proof that the trap is performing its functions correctly, and, after having discharged, the valve should close tightly and promptly at exactly the moment live steam appears. Dribbling also prevents assurance that the full quantity of water is being discharged, so that under these circumstances it is only natural to imply that water of condensation is backing up the inlet pipe, which may ultimately spell disaster to the system being drained.

From what has been said, it will be recognised that one feature of a good trap is that the valve should open wide, and remain so as long as water is passing, but immediately steam appears it should shut with a snap. Both dribbling and wire drawing cause rapid deterioration of the valve faces, with consequent inefficiency due to leakage of steam. A glance at figs. 2 and 3 will explain this point far more emphatically than any amount of words, for in fig. 2 we see a trap under steam at 180 lb. per square inch, with no sign whatever of dribbling or of a leaky valve, whereas in fig. 3—which was taken only a moment after—we see the same trap discharging its contents at a rush, after which, immediately live steam

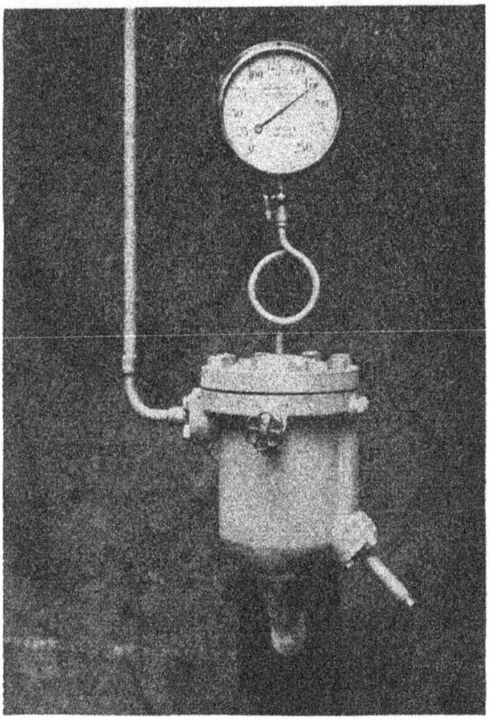

Fig. 2.

appeared, it again assumed the dormant appearance of fig. 2. It continued working so with almost rhythmetic precision. Such a discharge as this is very proof itself that the trap in the photograph is performing well. As will be recognised, it is of the bucket type, known as the "Sentinel," further particulars of which appear on page 55.

Discharge should take place as near as possible at 212 deg. Fah., for it must be remembered that a trap, to be efficient in the full sense of the word, should at the same time be economical as regards its consumption of heat. Of course there are cases in which this is not practicable, as, for instance, in the draining of the steam jackets of engine cylinders, for in such a case as this it is better to

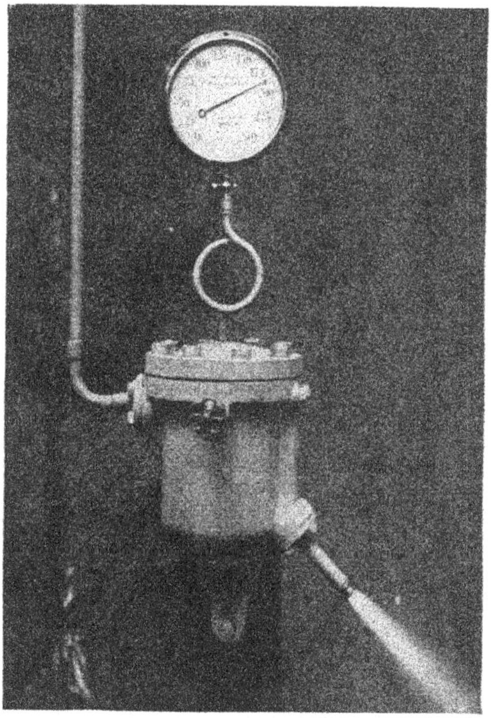

FIG. 3.

lose a little heat for the sake of greater economy in other directions; this may also apply to coils for boiling liquids, where it is imperative that the water should be discharged immediately it forms, so that an almost continual discharge of wet steam, although scouring the valve and its seating, is in this case a necessary evil. It should be remembered, however, that this scouring could never be very severe, as only a very low pressure would be used. For very low-pressure work a trap with a full bore right through should be selected.

Accessibility of the working parts is a feature which should be looked for, as it should not be necessary to break joints in order to inspect or renew a valve, or to make some trivial internal adjustment. A good trap for general engineering work should have a wide range of working pressure; the vital parts should be strong and not liable to corrosion, thereby ensuring, to some extent, a small outlay for maintenance in the renewal of parts. The valve-operating levers, etc., should be relieved of strain as much as possible, as otherwise it would be well-nigh impossible to obtain anything like a sensitive action. When the valve is opened against steam pressure it should always be relieved by a spring or some other suitable device.

To fulfil its duties properly, a trap, beside being very sensitive, should also remain wide open when cold, thereby enabling the steam pressure, when first turned on, to expel all accumulations of water and air from the pipes; otherwise very serious water hammer and rents may occur, due to the accumulation of cold water suddenly receiving the impact of live steam—the true cause of many an accident to life and property.

Should it be necessary for a trap to discharge above its own level, as in mine or marine work, an allowance of 2 ft. per pound of steam pressure is a safe working figure; in such a case a non-return valve should be fitted immediately on the discharge side of the trap. In laundry work steam traps are often arranged so as to return the water of condensation to a tank situated somewhere above the ground level, the water being used again for other purposes. An ingenious arrangement of trap and discharge, whereby water is again returned to the boiler against steam pressure, is described on page 83.

Summing up briefly, the functions of a good steam trap are threefold: (1) The removal of all condensation water from any point in the system; (2) to do so without wasting any steam; and (3) to do so without requiring constant attention at the hands of the engineer in charge.

It is no easy matter to lay down hard-and-fast rules for the classification of traps. Undoubtedly the best plan is to base their classification upon the principles of their working, and this is the method I propose to adopt here, as follows:—

Those which depend for their working upon—

Class 1.—The differential expansion of metal rods or tubes, or a combination of both.

Class 2.—The action of a float or bucket operating a valve.

THE THERMOSTATIC PRINCIPLE.

Class 3.—The action of the Bourdon tube.

Class 4.—The action upon a corrugated metallic chamber containing a volatile fluid.

Class 5.—Differential water pressure.

Class 6.—Return-feed systems.

CHAPTER II.

EXPANSION STEAM TRAPS.

Class 1.—**Those which depend for their working upon the differential expansion or contraction of metal rods or tubes, or a combination of both.**

It is a well-known fact that all metals expand or contract due to variations of temperature; these expansions are both cubical and lineal, though it is the latter we have to deal with here.

Take, for instance, a brass rod, and heat it in a flame. You will see no appreciable increase in its length, although an actual increase has taken place, and were you able to measure this increase with a fair degree of accuracy by methods already familiar to even the youngest physicists, you would find that the actual increase was directly proportional to the rise of temperature of the bar, and also, upon cooling down again, the brass bar would obey precisely the same physical laws as regards its contraction, finally assuming its initial condition. With the methods resorted to in order to measure these actual increases or decreases we need not enter, sufficient to say is that by more or less elaborate instruments it is possible to prove that the increase in linear expansion or contraction is directly proportional to the rise or fall of temperature, from which we obtain for every specimen of metal a figure known as the coefficient of linear expansion, which expresses the actual increase or decrease per unit length per degree rise or fall of temperature.

The discovery by a French chemist, M. Guilleaume, of an alloy of nickel and steel 36 per cent nickel and 64 per cent steel, provided us with a metal alloy having an extremely low coefficient of expansion, which, from the table following, is seen to be about ten times less than that of iron. This special alloy, beside being used to a great extent for instruments of precision, has been employed in steam-trap design, the most notable instance being in the trap known as Granger's, a description of which will follow shortly.

The following is a list of the coefficients of linear expansion of some of the most important metals used in the manufacture of steam traps:—

Iron	0·000013
Copper	0·000017
Brass	0·000018
Steel	0·000012
Nickel steel	0·00000133

From this list we see that those early designers who employed differential expansion as the principle upon which their traps worked would naturally choose the two metals having the widest difference in their coefficients, provided they could at the same time retain their tensile strength under moderately high temperatures; that is, he would choose steel and brass, the coefficients of expansion of which are 0·000012 and 0·000018 respectively, or in the ratio 7 to 10 nearly, where 70 per cent of the actual expansion would be mutual, leaving but 30 per cent for operating the discharge valve mechanism.

The idea of employing differential expansion—that is,

the rate of increase of difference of expansion—has provided a type of trap possessing several extremely interesting features, chief among which are its simplicity and the irresistibility of the expansion. Thermal expansion, however, although affording an irresistible force—limited only by the strength of the material used—for operating the valve, is very small, unless rods or tubes of proportionate length or levers for multiplying their expansion are employed. This point has been very successfully coped with by designers, as the following descriptions will make evident.

The following eight traps are typical examples illustrating the principles of Class 1:—

(1) Messrs. Holden and Brooke's trap.
(2) Messrs. Ogden's annular trap.
(3) Mr. W. A. Granger's trap.
(4) Messrs. Geipel and Lange's trap.
(5) Messrs. Boyles' trap.
(6) The "Columbia" trap.
(7) The "Reliance" trap.
(8) The "Arctos" trap.

It should be noted here that whereas some writers on steam traps would not consider at least three of the above traps as coming under the heading of "differential expansion," preferring to call it "controlled expansion," the writer can see but such a slight difference in the actual meaning of these two terms when applied to steam traps as to sufficiently justify him to classify them all under one head.

We will now describe the *modus operandi* of the above eight traps, drawing particular attention to any outstanding features.

Messrs. Holden and Brooke's (1903 Pattern) Trap.

This trap, a section of which is shown in fig. 4, has many interesting and original features, and on the whole is very efficient and reliable. It consists mainly of a hollow tube A, having at one end a valve-box arrangement B, and at the other end a rocking piece J, the two parts B and J being connected by two solid rods R and R^1. The

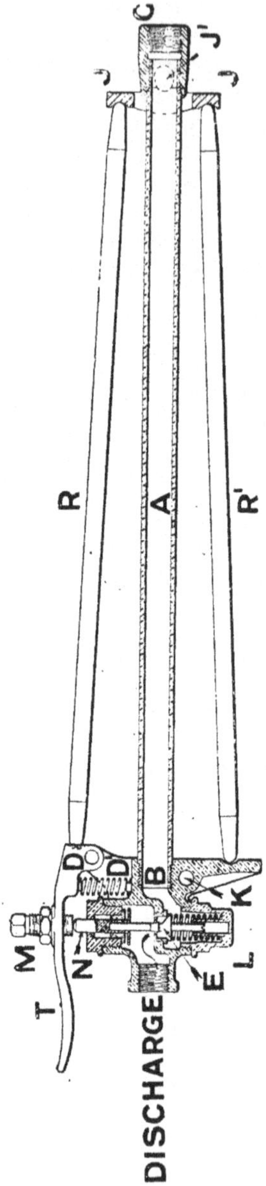

FIG. 4.—The Holden and Brooke (1903 Pattern) Trap

action of the trap is as follows: When the tube A is full of live steam it is expanded, holding the two parts B and J apart. Immediately water of condensation appears, being cooler than the steam, the tube A contracts, drawing the two ends together. This action tends to compress the two solid rods R and R^1. Thus the upper rod R presses on the lever D, which actuates the valve E and permits the water to escape. The rod R^1, acting on the rocking piece J, adds its own relative motion to that of R, thereby practically doubling the movement of the lever D. When

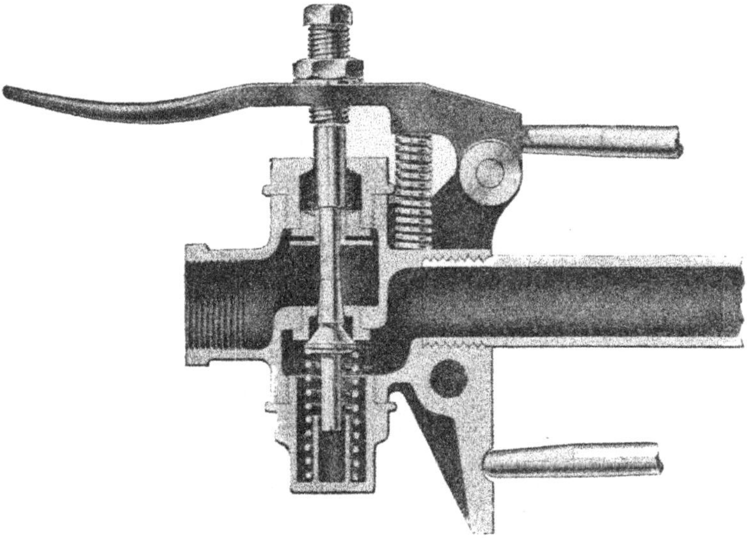

FIG. 5.—Valve of the Holden and Brooke (1903 Pattern) Trap

discharge has taken place and live steam enters the trap, the tube A expands, thereby releasing the pressure on the lever D caused by the rods R and R^1, thus allowing it to rise by virtue of the spring D and the steam pressure behind the valve, which re-seats itself instantly. **M** is an adjustable screw and lock nut used for regulating the discharge, and T is a lever, which, upon being depressed by hand, allows steam to blow through. The main function of the spring B is to always keep the rod R in the little cup formed for it in the lever D.

One of the most interesting features is the shape of the valve, which is best seen on reference to fig. 5, an enlarged sectional view of the valve head. This valve has been designed with a view to prevent dribbling, for, as will be seen, it is provided with a cone or extension, which, when closed, nearly fits a similarly-coned hole above the valve seating proper. Owing to its slower angle, this extension opens very much slower than the main valve, and thereby practically prevents the trap from discharging until the valve proper has been moved sufficiently off its seating to prevent dribbling or cutting of the valve face; also the fact of the valve having to be forced off its seating against steam pressure; it requires a somewhat greater force to commence to open it than is required once it has moved, as, immediately the valve commences to open, the pressure on it would decrease, allowing the lever D to suddenly overcome the resistance and open the valve to its full extent, so that by this means also dribbling is guarded against, and the action of the valve becomes a positive "stop" and "start" action.

For a trap working upon the principle of expansion it has a very wide range of working pressure without requiring adjustment. The makers claim this to be about 120 lb.—that is, from 5 lb. to 125 lb. say. The writer has never had the opportunity of testing this within such wide limits. The trap is very sensitive to small differences of temperature, a difference of but 7 deg. being sufficient to cause it to blow, for it may be made to instantly discharge by pouring a little water on the tube A.

The valve is easily renewed or inspected by unscrewing the cap L, fig. 4, and, as well as possessing but few working parts, the valve always remains wide open when cold, which, it will be remembered, is a very desirable feature. In common with most traps of this class, it works well in almost any position, and is capable of delivering against a head.

Ogden's Annular Steam Trap.

Fig. 6 is a section of Ogden's annular steam trap, and, as will be readily seen, its design is very similar to that of the trap just described, for it embodies precisely the same principles; there is the brass tube and the valve operating rods and lever applied in a similar manner. However, the trap contains several characteristic points, which cannot fail to be of interest, chief among which is the valve, which is of a decidedly unique design. The slightest mechanical movement—due to the contraction of the central tube—being transmitted to the valve is sufficient to cause it to open slightly and allow a little water to pass, which commences the discharge, with the result that the escaping water, passing the annular faces on the valve—clearly seen in fig. 6—at a somewhat high velocity, causes it to pop open full bore, or many times wider than the rod and lever mechanism is capable of doing under fine conditions of working. Directly all the water is discharged, and as soon as steam attempts to pass, the spring load shuts the valve with a snap, the action being accelerated on account of the reduced momentum of the steam, which does not have the same action on the valve faces as the discharging water under pressure has. This independent movement of the valve gives the trap a much greater capacity for discharge than if the mechanical movement due to the tube contracting was the only valve opening medium relied upon.

Since the valve at each discharge opens practically full bore, an extremely vigorous discharge takes place; a valuable scavenging action is thereby brought into play, for condensed water on the sides of the pipes and vessels being drained, which would perhaps remain there, is carried along in the train of the discharge, instead of being left for a future and what would of necessity be an intermittent dribbling discharge. So that the discharge is marked by its intensity and volume, a point which almost entirely eliminates the possibility of dribbling.

Another valuable feature of interest is the iron rod, which may be seen inside the central tube. By its use an annular space is provided for the water in the tube. This has a double advantage, for it requires but a very small quantity of water to fill the trap, and so produces a quicker contraction; and also in the act of vigorously discharging it causes the discharging water to scour away any grease deposits from the inside surface of the tube.

which would otherwise accumulate and destroy the sensitiveness of the trap when working under dirty and greasy conditions.

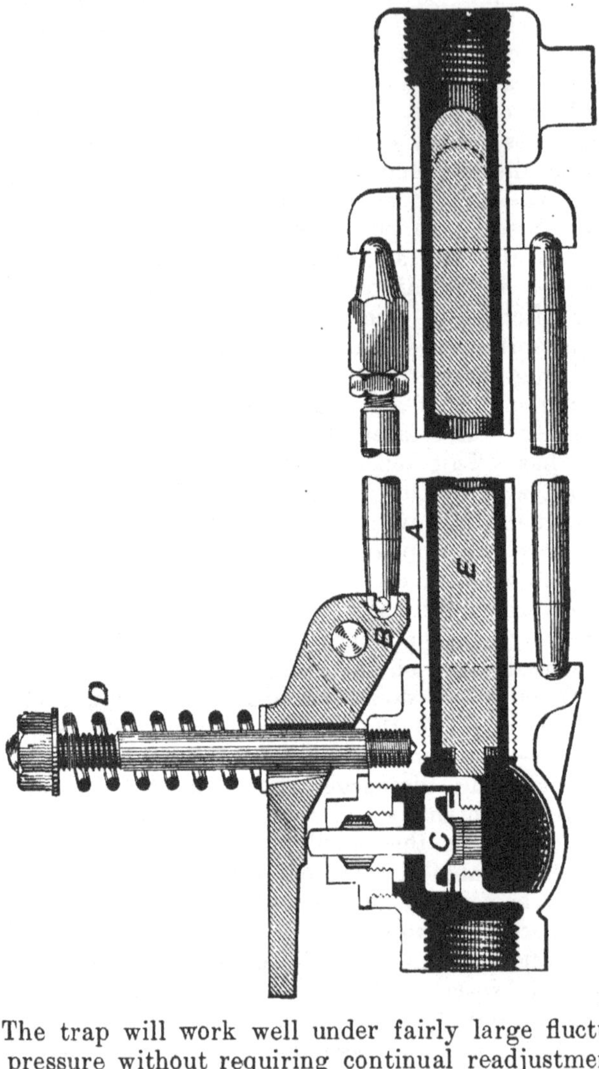

Fig. 6.- Ogden's Annular Trap.

The trap will work well under fairly large fluctuations of pressure without requiring continual readjustment, due to a peculiar compensating action between the fixed spring load and the variations of pressure. The parts are very accessible, as in the Holden and Brooke trap, and the valve is easily and quickly renewed, or it may be taken out when slightly worn and its flat surfaces ground up true upon a piece of emery paper placed on a flat board. Adjustment is rendered possible by varying the spring load, and also by means of the locking nuts arrangement on the left-hand end of the upper horizontal rod. A blow through is provided for in the usual manner.

Granger's Steam Trap.

This trap, as shown in fig. 7, deserves a much greater share of patronage than that which it already receives, for, in the opinion of the writer, the design is a distinct mechanical success, especially when employed with high-pressure steam mains. It consists of a brass tube K K, having gun-metal fittings at either end, and a bar A C of nickel steel fixed at A. As mentioned previously, this nickel-steel bar has an extremely low coefficient of expansion, so that the distance from A to B, the fulcrum of the lever L, remains practically constant through a very wide range of temperature. Now, when the body of the trap expands, the lug H, which is a loose fit, slides along the rod A C, transmitting its motion, by means of the stud R E, to the lever L, by which it is multiplied four or five times before reaching the valve V.

The normal position of the brass tube K K—that is, its position when full of live steam—is so that the valve V is hard on its seat; however, when water of condensation accumulates, it contracts, so that the lug H slides along the rod A C towards the left. In so doing its motion is transmitted to the lever L through the stud R E, at the same time being multiplied four or five times before reaching the valve stem at G. As soon as all the water present has been ejected, the tube K K fills with steam and expands, pushing the lever L to the right, and closing the valve.

A great feature of this trap is the balancing of the valve by means of the spring S, which will be seen better on reference to fig. 8. This spring counteracts fully three-quarters of the pressure on the back of the valve, which, in cases where high pressures are used, amounts to no mean trifle. Take, for instance, a trap with a $\frac{3}{4}$ in. valve, working with a pressure of, say, 200 lb. per square inch; then, if the valve be entirely unbalanced, the dead pressure

FIG. 7.—Granger's Trap.

upon it would be about 88 lb. This pressure would be transmitted to the stud R E, and owing to the leverage of the lever L the actual force acting on the stud R E would amount to nearly 440 lb. tensile. Under such conditions the trap would experience abnormal strains, and a sensitive action would be well nigh impossible. It is here that the value of the spring is felt, for, as mentioned, it relieves the valve of three-quarters of the pressure due to the steam; so that this enormous strain of 440 lb. acting on the stud R E is thereby reduced to something in the neighbourhood of 50 lb. or 60 lb., whereas the force required for opening the valve is not in the least diminished.

Dribbling is effectually obviated in the following manner: As soon as the valve has moved but slightly off

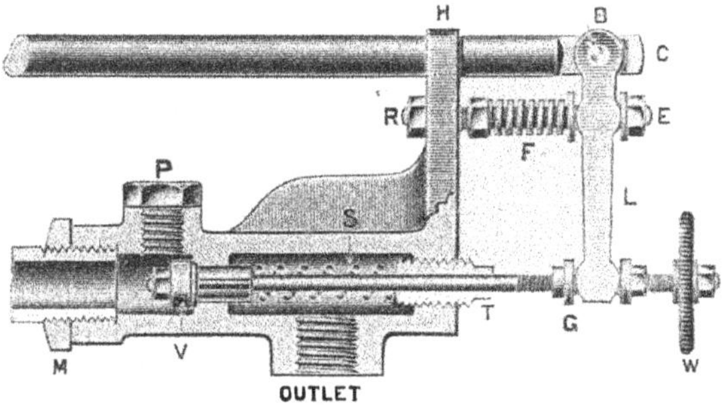

FIG. 8.—Valve of the Granger Trap, showing how the valve is balanced.

its seating, owing to the contraction of the brass tube K K, it then opens suddenly, partly on account of the relieved pressure on its face, and partly due to the action of the spring S and the slight elasticity of the nickel-steel rod A C. After the water has been ejected the subsequent expansion of K K, combined with the action of the spring S, suddenly draws the valve back on to its seat with a snap, so that the desirable form of discharge is obtained—namely, puffs.

As this type of trap depends for its successful working almost entirely upon temperature, and as also the temperature of steam varies with the pressure, it is obvious that Granger's trap and all other traps belonging to this class will not work so well when subjected to varying pressures. For instance, imagine a pressure fluctuating between, say, a maximum of 200 lb. and a minimum of 20 lb. in a few hours; it is obvious that under such conditions, which represents a difference of temperature varying as much as 150 deg., an expansion trap when set to work at the low pressure could not possibly deal effectually with the high pressure without having to be continually readjusted, a course which is practically impossible. However, Mr. Granger has introduced an additional improvement in the form of an automatic pressure-compensating gear for use with his trap, which very easily gets over this difficulty. This fitting is represented in fig. 7 by the dotted line Y N J, and is also seen enlarged in fig. 9. It consists of a plunger N working in a cylinder Y, the bottom of which is in direct communication with the steam pressure through the passage X, so that the piston, which is kept down by a spring, rises or falls according as the pressure fluctuates. As the piston rises, in so doing it—by means of the arm Q—gives a partial rotation to the stud J R E, fig. 7, for which the lug N then forms a nut (the nut R being absent when this additional gear is fitted). The effect of this partial rotation is to slightly move the stud longitudinally towards the left, thus pushing the lever L in the same direction, and thereby causing the valve to open somewhat later than it would under previous conditions. To illustrate this the following example will be useful: Suppose that the difference of pressure at any instant from that existing a few moments beforehand resulted in K K being subjected to a temperature at which discharge would take place of, say, 160 deg. rise, then the length of K K would be correspondingly increased by 0·000018 multiplied by the length of K K multiplied by 160—that is, about one-twentieth of an inch—so that the stud R E would be one-twentieth of an inch too far towards the right; but were a pressure-compensating gear fitted the piston N would rise to the position Q, fig. 9, and thus cause the stud R E to shift one-twentieth of an inch to the left, thereby bringing it back to its original relative position. This dimension is not strictly correct, since the valve seat itself would move slightly to the right, and consequently the backward movement of R E ought to be slightly less than one-twentieth of an inch.

To obtain access to the valve for the purpose of regrinding, it is only necessary to remove the plug P and insert the grinding powder, rotating the valve by means of the hand wheel W. When grinding in the valve, the lock nut E should be slacked out, so as to disengage the lever L. If it is necessary to remove the valve, the plug D at the further end may be taken out and the valve pushed through. In common with most traps of this class, the valve remains wide open when cold.

Messrs. Geipel and Lange's Steam Trap.

Of all traps working upon the principle of expansion, that patented by Mr. Wm. Geipel as far back as 1864 is at once the most ingenious and the simplest on the market, for he was the first designer to hit upon the exceedingly simple

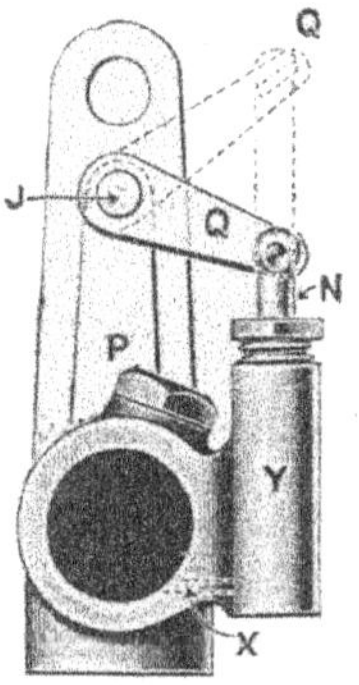

Fig. 9.—Automatic Pressure Compensating Gear on Granger's Trap.

idea of utilising the geometrical properties of the isosceles triangle whereby the differential expansion of the two sides, combined with the peculiar properties of the triangle, is employed to operate a valve.

Briefly, the system consists of two metal tubes of iron and brass respectively, so placed as to form the two sides of an isoscèles triangle, the base of which is a fixture, whilst the apex is free to move in a direction practically at right angles to the direction of the linear expansion of the tubes. By this means the actual difference of lineal expansion is multiplied about ten times, without the intervention of multiplying levers of any description.

Fig. 10 shows a part section of the trap. The lower or brass pipe constitutes the inlet, whereas the upper pipe, which is of iron, forms the outlet. These two pipes are placed in an open work-box frame, so as to form the two equal sides of the isosceles triangle, the valve D being the apex. When water of condensation is present in the brass inlet pipe, being cooler than the steam, the pipe contracts, and in so doing drags down the apex of the triangle, thereby displacing the valve and allowing the pressure to expel the accumulated water. As soon as steam appears the inlet pipe expands, raises the valve seat, and closes the valve.

For ordinary use on land, and for work where the steam pressure does not vary to any great extent, the ordinary

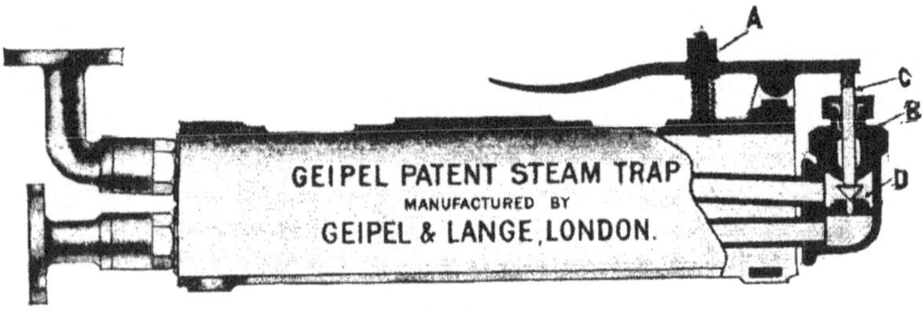

Fig. 10.– The Geipel Trap.

type of this trap holds its own, and receives a very fair share of patronage. But in cases where the pressure is liable to tremendous fluctuations, as in marine work on battleships, etc., in order to comply with the exacting requirements of the British Admiralty, where the pressure may be anything from that of the atmosphere to 300 lb. per square inch, or over, Mr. Wm. Geipel set himself, with great success, to design the Admiralty type Geipel steam trap, which complies exceedingly well with the apparently anomalous conditions thereby entailed—namely, to take extreme cases, the Admiralty require a trap the valve of which is to remain tight at every steam pressure from atmospheric pressure to 300 lb. per square inch, and at the same time to discharge water as quickly as it is formed; that is to say, it must hold steam at a temperature of 220 deg. Fah., and yet discharge water at 410 deg. Fah. Obviously the ordinary type of Geipel trap alone would be out of the question for such use; but by means of a very simple pressure-compensating fitting, whereby the varying steam pressure acting upon a flexible diaphragm renders the trap automatically self-adjusting. Fig. 11 shows a Geipel trap of the Admiralty type, and it is a sufficient testimony in itself of its efficiency to state that the British Admiralty have adopted this trap almost

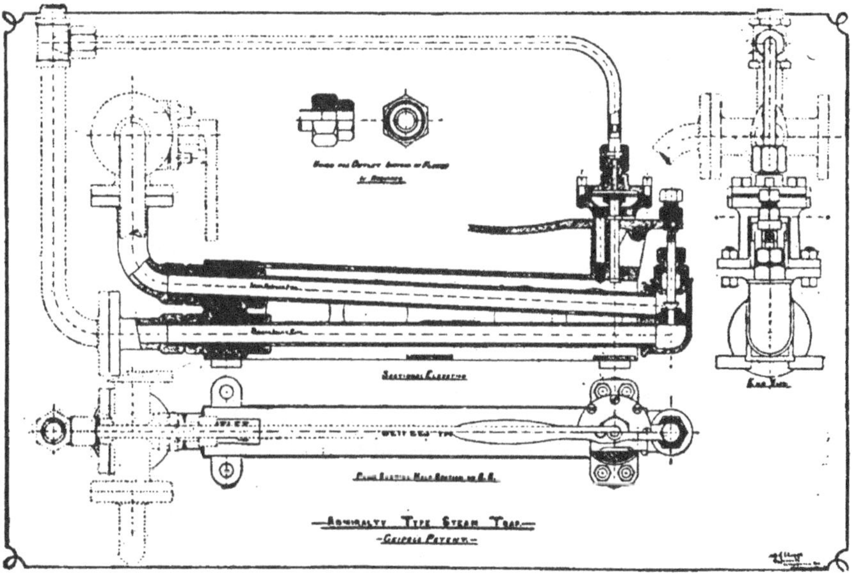

FIG. 11.—Geipel's Admiralty-type Trap.

exclusively as being the one which best complies with the rigorous conditions—inseparable from Government specifications—under which a steam trap could possibly work. The action of the expansion members is the same as in the ordinary type, but to overcome the pressure-fluctuating difficulty the hand adjustment by means of the nuts A, fig. 10, is replaced by a collar, which is operated on by a diaphragm subjected to the pressure in the steam pipes. As the movement of the valve seat at the apex increases as the temperature increases, adjustment is obtained in the ordinary type trap by screwing down the nuts A for high-pressure regulation and slackening them for lower pressures. Now, as will be seen on reference to fig. 11, the varying pressure in the main steam pipe is transmitted to the diaphragm chamber through a by-pass, so that the collar, being controlled by the motion of the diaphragm, depresses the lever when the pressure rises, and allows it to rise when the pressure falls.

As an instance where the Geipel Admiralty trap has been adopted with marked success, we may cite the case of steamship Repton, whose engine power was increased from 847 to 903 horse power. Similar results were obtained by Professor Weighton at Newcastle-on-Tyne. He found that the use of these traps increased the power of an engine by 5 per cent, and at the same time reduced the steam consumption per hour by 5 per cent.

Among its good points, the valve is easily inspected or renewed without breaking joints, the valve is wide open when cold, and a direct blow through may be obtained by depressing the lever by hand, an extension being supplied for the purpose. Mr. Geipel, regarding expansion steam traps, noted that the valves did not always close in a steam-tight manner after having been in use for some time, because the wear on the seat was not always uniform. The axis of the valve being usually at right angles to that of the expansion member, the expansion not only opens the valve, but puts side pressure on the spindle. He therefore took out a patent for an expansion trap in which the valve spindle is placed in an inclined position, so as to prevent the possibility of this side pressure.

The method of carrying this into effect is seen in fig. 12. The valve case *a*, valve *e*, and spindle *f* are inclined to the centre line *x x* between the expansion tubes *b c*, and to the length of the trap from *d*. The arm *g* of the spring-controlled hand lever *g h* is also set so that the valve spindle normally abuts upon it at right angles, and there is then practically no motion between the contact surfaces when the trap opens and closes, this overcoming all tendency to produce a tilting action on the valve *e*. For a trap of the size given in the illustration the angle of the line of the valve spindle to the centre line of the tube *b* is about 64 deg.

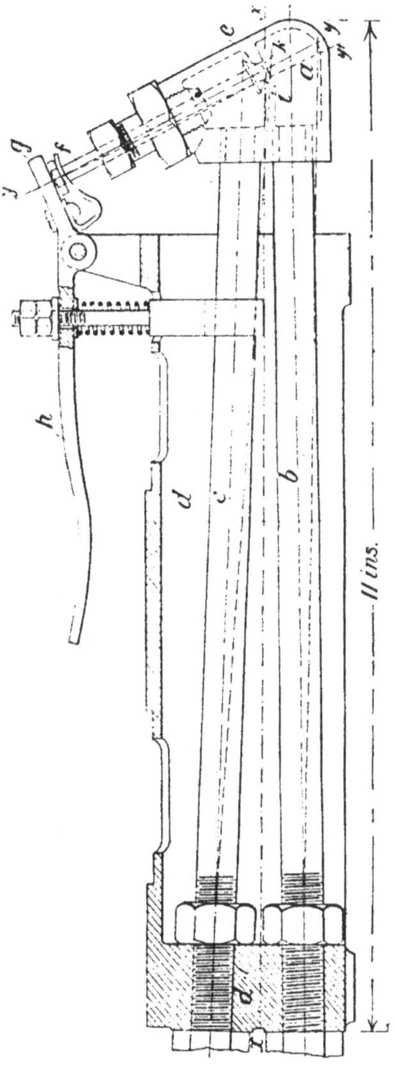

FIG. 12.—Geipel's Improved Trap, with Inclined Valve Spindle to Prevent Side Pressure

To ensure a uniform wear of the discharge valve and its seat, Mr. Geipel takes advantage of the relative motion of the valve case to the trap frame. By means of a bell-crank lever with a pawl affixed to it at its upper end a slight turning movement is given to the valve each time an expansion and contraction takes place.

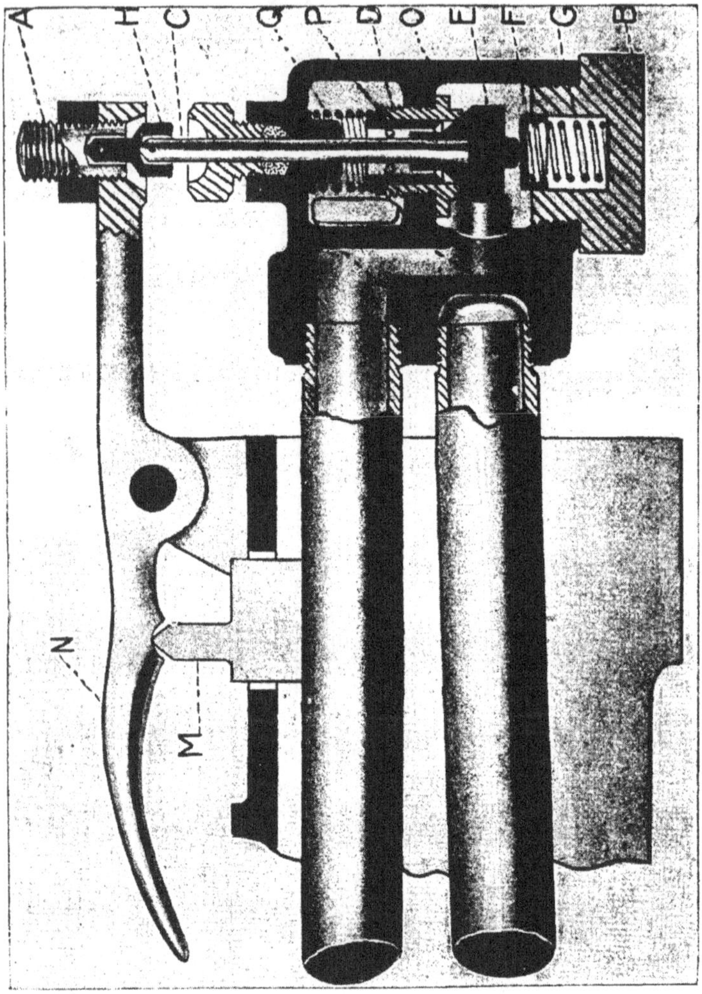

Fig. 13.

Mr. Geipel's latest improvements are always of interest, for his experience in designing steam traps has made his name familiar to all those interested in the subject. His very latest improvement consists of a thorough re-arrangement of the discharge valve mechanism of the ordinary type of trap bearing his name. He has called this new type the "Rapidity" Geipel steam trap, and claims the following improvements:—

(1) a larger valve area, and consequently a more perfect discharge;
(2) A means whereby the valve tends to re-grind its face at each discharge; and
(3) A means whereby the noise, strain, and other deleterious effects of a too rapid discharge at high pressure is decreased by an automatic throttling device.

How far the addition of these improvements has altered the general appearance of the discharge mechanism of a standard Geipel may be judged by reference to fig. 13, in which it will be seen that the converse arrangement of the brass and iron tubes is employed; the brass inlet pipe now being situated at the top. This arrangement permits of a much larger valve area being employed, for, as shown, the valve is arranged in such a manner that it is held on its seat by the steam pressure, which also is the converse arrangement, for in previous types the valve closed against steam pressure, which had to be overcome before the valve could open for a discharge.

When the pressure behind a discharge is considerable a great deal of noise, strain, and cutting action is set up, which is bound to have a most deleterious effect upon the trap as time goes on. To destroy this action without impairing the discharge capacity when working at low pressures has always presented a problem of some difficulty, the usual practice having been to choose upon a discharge valve area which would be calculated to give a fairly good all-round efficiency when working with any steam pressure. This method, whilst serviceable, is by no means perfect. To overcome the difficulty Mr. Geipel has fitted an automatic throttling device, which comes into action at high pressures, but not under low, so that a full-bore discharge at low pressures and a controlled discharge at high pressures are obtained.

As seen in fig. 13, the valve E is held on to its seat by the steam pressure combined with that of the light spring G. When the body of the trap is cold the valve is held off its seat by the spindle C, which is held in position by the blow-through lever N. But when the upper or brass tube is expanded, due to the presence of live steam, the valve being subjected to the steam pressure is held hard on its seat until water of condensation has accumulated in a sufficient quantity to cause the brass inlet pipe to contract, which action, being exactly similar to that of the ordinary type of Geipel, opens the valve and causes a sudden rush of water over the valve face. This rush is instrumental in bringing into action two very useful features; for, firstly, the valve E, being loose on its spindle, is forced or wedged down against the action of the small spring G, situated immediately beneath it; this action very greatly increases the area of the discharge passage. Also the valve, being provided with vanes, is rapidly rotated, which gives a tendency to re-grind the valve at each discharge.

In the case of a discharge taking place at high pressure, in order to overcome the disadvantages due to the noise, strain, and other deleterious actions which would be brought into play, the automatic throttling device, above referred to, comes into action. This device consists of a movable bush P, controlled by a spring Q, which is forced up against the action of the spring by the rush or momentum of the discharge, thereby reducing the area of the passage, and consequently the velocity of the discharge. Since the height which the bush P would be lifted depends upon the momentum of the discharge, it is obvious that this throttling device can be set so as not to come into action at low pressures. On account of the sharp blow-through which occurs at discharge, together with the tendency of the valve to remain well ground to its seat, it is claimed by Messrs. Geipel and Lange that this new type Geipel is remarkably free from dribbling.

MESSRS. ROYLES' EXPANSION STEAM TRAPS.

The traps manufactured by Messrs. Royles Limited are practically the simplest designs of the expansion types now on the market. The illustration, fig. 14, is a section of one which almost explains itself. It consists principally of a gun-metal tube, having at its centre a valve box A contain-

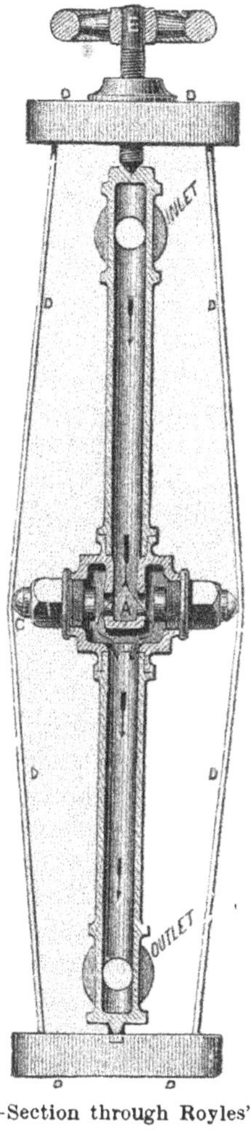

FIG. 14.—Section through Royles' Trap.

ing two flat-faced valves C C, the latter being held in place by two bow strings D D. These bow strings are rigidly cast into the head and tail pieces D D. Obviously, should the gun-metal tube expand, the head and tail pieces would move in opposite directions, thereby causing the bow strings to move inwards at their centre, and exert a pressure on the valve stems at C. The discharge is adjusted by means of the hand wheel E, on the top, so that when live steam is present the valves are hard on. Now, any slight contraction of the main tube would slacken the bow strings, and enable the steam pressure to force the valves outward when the discharge takes place. The trap is very sensitive, and has a fairly wide range of working pressure. It may be easily blown through at any time by manipulating the hand wheel, the only disadvantage, however, in so doing being that the adjustment is thereby upset; but since adjustment is so easily and quickly effected, this is not a point of any material importance.

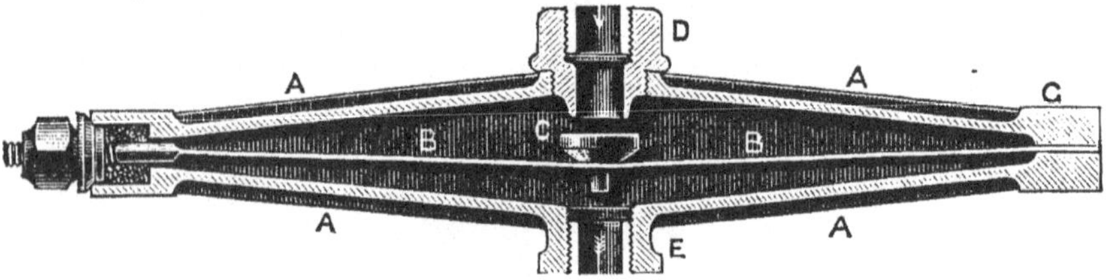

FIG. 15.—Royles' 1893-type Trap.

Another trap on much the same lines, but of a much more compact form, is the 1893 pattern of Messrs. Royles' trap, as illustrated in fig. 15, where it will be seen that the same bow-string principle is adopted, but in a modified form. The body A of the trap is of gun-metal, and the bow-string bar B is of steel. The latter is firmly cast in position by using it as a core bar when casting the body B. As will be seen, the other end of the bar passes out through a packed gland, being fitted with an adjustment nut on the outside. The bar B is slightly bowed, and carries at its centre the flat-faced valve C. Water of condensation enters through the inlet tube D, flows round the bow string, and out at the outlet E. Immediately live steam appears, the body of the trap fills up, and consequently expands; the tension thereby induced in the bow string pulls the valve hard up on to its seat.

The simplicity and low initial cost of this trap renders it especially valuable for use with steam-heating installations, as a separate trap may be set to work at each drip without running up the cost unduly.

Another design of trap of this class recently placed on the market by Messrs. J. and *C.* Hayton, of Bradford, and known as the "Arctos," possesses great simplicity and compactness, and has an originality in design which cannot fail to attract attention. It is constructed on the expansion principle, but, unlike most expansion traps, it is fitted with two expanding pieces of like metal, and the principal object sought has been to obtain in a small-size trap the greatest possible valve lift, whilst obtaining a

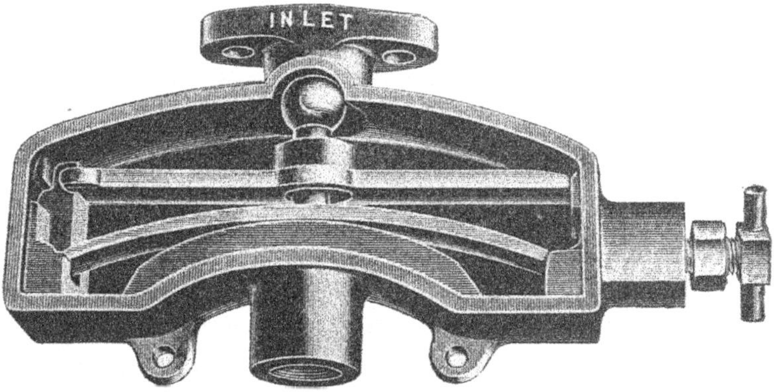

Fig. 16.—The "Arctos" Steam Trap.

maximum of strength and avoiding the employment of numerous accessory parts.

A reference to the illustration will serve to show that the valve lift must be considerably greater than the extent to which the bow piece on which it is supported expands; for, in the first place, the deflection of the bow at its centre will be about three times the extension due to expansion, and the extension of the second expanding piece will cause a further deflection; and, secondly, by the lever or toggle arrangement shown the deflection can be further considerably increased. In this manner the expanding bow, which is only about 9 in. in length, can provide a valve lift equal to a direct expanding tube or rod from 6 ft. to 9 ft. in length.

The trap has practically no pressure to withstand, as the valve is placed in the inlet and only opened for the escape of water. The position of the valve relative to the seat is regulated by the thumb screw, which deflects or releases the bow piece as desired. As the expanding pieces are not gripped or held immovable at either of their ends, any movement under expansion does not weaken their stability or gradually prevent them from regaining their original shape and position. The body of the trap is of cast iron and all the internal parts are of gun-metal.

Of American traps coming under the classification of those which depend for their working upon the expansion

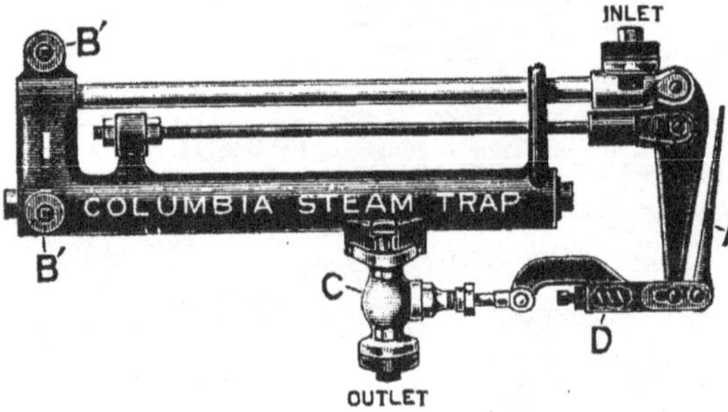

FIG. 17.—The Columbia Steam Trap.

of metal rods or tubes, that known as the "Columbia," and manufactured by Messrs. Watson and McDaniel, of Philadelphia, is the only one of any importance. An outside view is shown in fig. 17. Its construction consists of four essential parts, namely: A brass tube, which, being subjected to varying temperatures, exerts the irresistible force of its expansion and contraction to operate a valve; a chamber in which the water of condensation may accumulate after passing through this brass tube; an iron rod, the length of which is adjustable, controlling the motion of the valve by means of a lever suitably pivoted to the expansion member and acting about the end of this rod as a fulcrum, and a valve box with a suitable discharge valve.

The water flowing in at the top point, marked "Inlet," passes through the brass tube into the cast-iron accumulator, which forms the body; and in common with other types of expansion traps, so long as water is passing in a sufficient quantity the valve remains open, but directly live steam appears at the inlet the expansion member rapidly expands. Now, being fixed to the body of the trap at the point farthest from the inlet, it is obvious that it (the brass tube) must expand in the direction from left to right, as seen when looking at fig. 17. This action brings the valve-operating parts into play; for the lever connecting the expansion member to the valve stem is canted about the end of the centrally-situated iron rod as a fulcrum; this action obviously multiplies the amount of expansion about four times at the other end of the lever, which action closes the valve.

The spring at D is a safety device for the protection of the parts from the enormous strain which would be set up in the event of a considerable increase in working temperature. For if the valve was set so as to close before the expansion member had fully expanded, any further expansion would only cause the lever to compress this spring, and thus obviate any of the undue stress or buckling strain which the trap would otherwise experience.

The lever A is a very handy means of blowing through the valve. It is also obvious that, owing to the bulk of water in the iron body of the trap, between the expansion member and the outlet, that the valve is well protected, and the possibility of dribbling and the wasting of steam is reduced to a minimum.

A new expansion trap of British manufacture, and one which has not been upon the market very long, is that shown in fig. 18. It is very similar in appearance to that of Messrs. Royles' design; it entails many points of interest, and on the whole bids fair to prove quite a reliable and simple trap, and one which well merits a description, if only on account of the unique and original method of so arranging that the strain due to the working pressure and that due to the irresistible force of expansion work, as it were, against each other, thereby tending to maintain a state of equilibrium, and also making the system automatically compensate for mode-

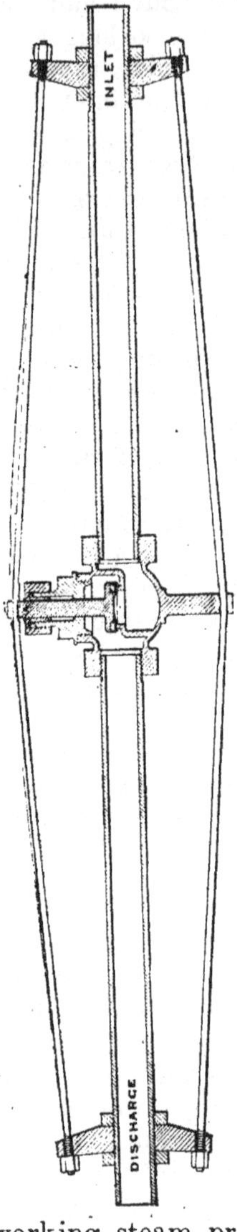

FIG. 18.—Section of the "Reliance" Trap.

rate variations of working steam pressure. Its construction consists of a solid-drawn brass tube, screwed at either end to take the end pieces to which the controlling members are attached; these members consist of a pair of tinned-steel wires, varying from $\frac{1}{8}$ in. in the smaller to $^3/_{16}$ in. in the larger sizes of the trap, there being six standard sizes. The central part of the trap is a strong bronze casting, and contains the discharge valve. The action, when in use, depends upon the alternate expansion and contraction of the central brass tube; the tinned wires performing a similar function to that of the bow strings in Boyles' design. An interesting feature is that the pressure and the temperature work against each other and tend to balance; for, upon observing the point of inlet to the trap on the right-hand side, it will be seen that the pressure in the system being drained is constantly tending to open the valve, whereas, when properly adjusted, the expansive force is exerted by the central tube; which force, transmitted to the valve through the agency of the tinned wires, has a corresponding tendency to keep the valve closed; this being so, it is obvious that a reduction in pressure, together with a corresponding decrease in temperature, would not upset the adjustment of the wires, provided the drop in pressure was not considerable; for, together with the balancing effect mentioned, the tension of the wires—which, being of steel, admit of flexibility—also tend to decrease proportionally with the pressure of the steam acting against the valve; but should the pressure rise much above that at which adjustment was made, the extra expansion in the tube would be absorbed in the elasticity of the wires, and a greatly increased inward pressure would hold the valve tight against this increased steam pressure. It is claimed by the makers—the Reliance Engineering Company, of Woodley, near Manchester—that the parts are so proportioned that the amount of contraction of the central tube is multiplied twenty-five imes at the valve seat; and this ensures a wide discharge opening, which, as previously mentioned, tends to keep the valve and its seat in good condition, and prevents dribbling.

CHAPTER III.

Float-operated Steam Traps.

Class 2.—**Those which depend for their working upon the action of a float or bucket operating a valve.**

The very simplest type of float-operated steam trap is that shown in fig. 19. It consists of merely a float and box; the action of the float in rising uncovers an opening in the bottom of the box, through which the accumulated water is expelled. This is a very simple arrangement, and were it not for the fact that it is essential to employ a very large float to obtain anything like a reasonable size of valve opening, the idea would be more generally employed, especially under low-pressure conditions, if only on account of its extreme simplicity. But simplicity must go hand in hand with reliability, otherwise a system would be doomed to failure.

The inlet being at the top of the trap, the float is subjected to the full pressure of the steam in the apparatus being drained, which pressure would be depended upon to hold the valve hard on in the presence of live steam; however, when the water accumulates, before the valve can open, the buoyant effect of the float must overcome the pressure acting upon the area of the valve opening, which, to cite a case of a trap working under, say, 120 lb. pressure per square inch, and having a valve, say, $\frac{1}{4}$ in. diameter, would be about 6·5 lb., which, together with the weight of the float itself, would have to be overcome by this buoyant effect before a discharge could take place.

Now, this difficulty might be got over by employing a large float; but this would be of little advantage unless we could also increase the area of the discharge opening. A little thought, however, will soon convince one that such an idea would only be practicable in conjunction with a very large float, which would be quite out of proportion, and at the same time would necessitate a somewhat expensive box.

A move in the right direction is shown by the illustration of the simple type of float arrangement in fig. 20. In

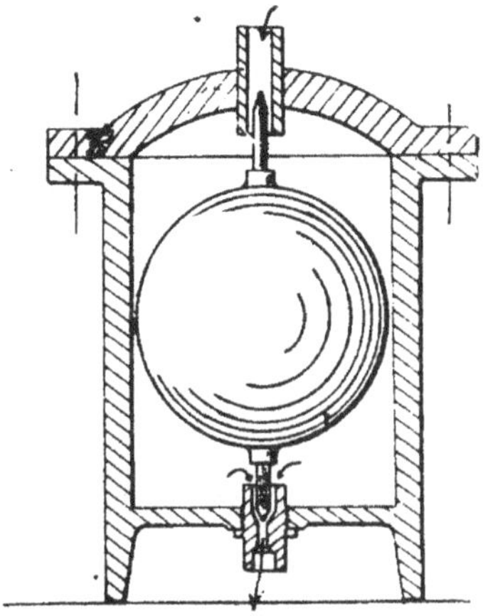

FIG. 19.—Simple Float-type Trap.

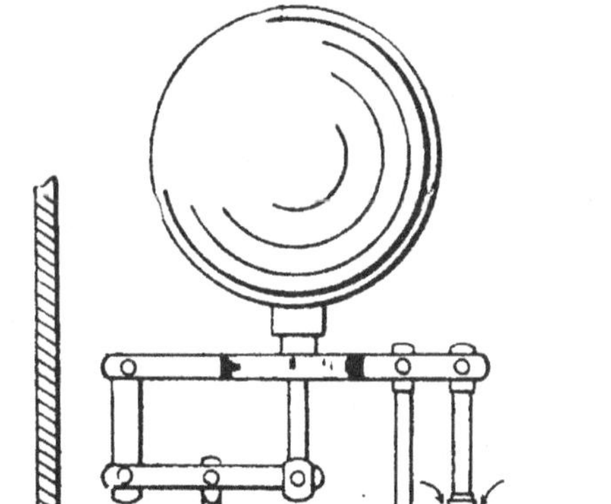

FIG. 20.—System of Multiplied Float Effort adopted in the "Meulleur" Trap.

this arrangement the upward force exerted by the float, when the body of the trap is full of water, is multiplied by a system of levers, in such a manner that the force lifting the valve is about five times that exerted by the float. This arrangement has proved successful, even with moderately high pressure. A point to be noticed is that the valve would be protected, on account of the float being in such a position that the valve would be closed before the body was completely empty, thereby protecting the valve under a water seal.

Another method employing levers for the multiplication of the float effort is shown in fig. 21, which is the method adopted by the Davies Regulator Company, of Chicago, and by which means a comparatively large valve opening may be obtained in conjunction with a small float. On either side of the rocking arm are pivoted the discharge

FIG. 21.

valves, the one opening inward and the other outward, which, being of the same size, are therefore balanced for steam pressure. Hence the force exerted by the float when the body of the trap is full need only be sufficient to overcome the small amount of friction at the pin joints of the lever.

Fig. 22 shows another method of obtaining a multiplication of float effort. This also employs a balanced valve. This method is adopted by the John Davies Company, of Chicago. Again, fig. 23, which is a section of the McDaniel improved steam trap, shows a still simpler, and at the same time a much less expensive, method of obtaining a multiplied float effort at the valve spindle.

For simplicity of design and freedom from complication the float or bucket type of trap has few rivals. It is true that their valve-closing forces are by no means so powerful as those of an expansion trap; but when the reader has studied this class he will see that such powerful forces are not by any means so important a desideratum here as they are in expansion traps. In fact, in the majority of cases the forces are quite powerful enough without being too much so, for it must not be lost sight of that where very

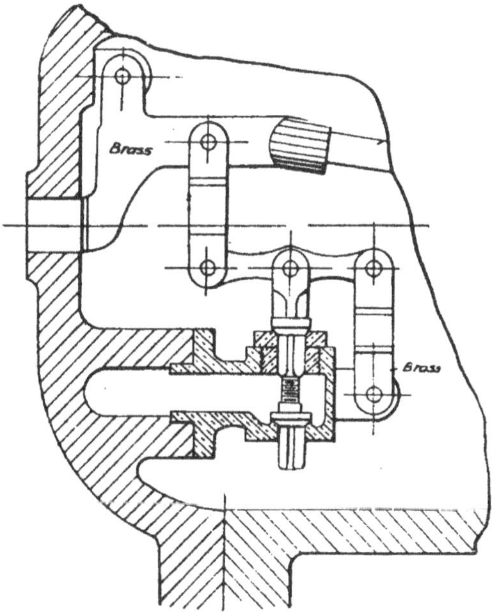

FIG. 22.—Valve of the Eclipse Trap.

big forces are brought into play the working parts must be designed to cope with them without fear of failure due to overdue strain or buckling. But with the amount of thought expended upon this point the valve-closing and operating devices are every bit as good and sometimes better than those of other types of traps.

From the nature of its design a float or bucket trap is not quite so efficient as are expansion traps, for the very principle involves the continual condensation of extra steam in order to operate the float. The above statement may also be easily demonstrated by placing the hand on the body of the trap whilst working. A great drawback is the fact that the body of the trap must always be full, or nearly full, of water, which is very liable to freeze when not in use, especially when the trap is placed in an exposed position, a fact which would not make itself evident until actual damage had been done.

One often hears it stated that traps of this class are liable to explode, or, rather, burst. Such accidents, unhappily, have occurred, but never can it be said that the trap was entirely to blame. Perhaps if men, when

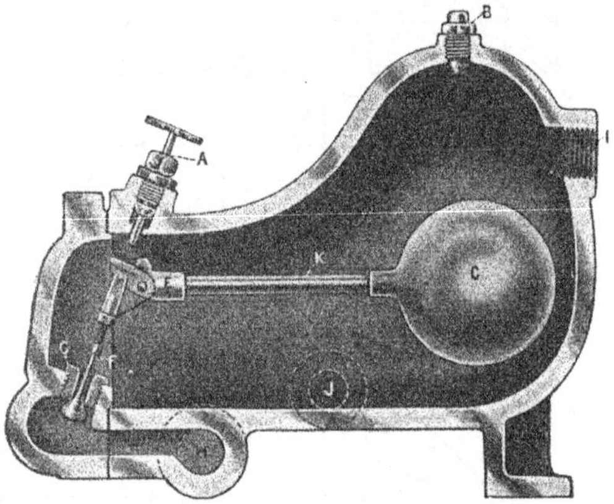

FIG. 23. - Section of the McDaniel Improved Trap.

placed in charge of traps, were thoroughly instructed and intelligently comprehended their principles, instead of looking upon a trap as a fitting capable of taking care of itself indefinitely, without as much as an occasional blow through, there would undoubtedly be much less undeserved scandal laid at the doors of bucket and float traps than is the custom of the salesman when dealing in other types.

They are almost entirely independent of boiler pressure, and entirely so from variations of temperature. Hence it is not necessary to fit pressure compensating gears or the like, a fact which makes every individual trap capable of a very wide range of adaption. They may be set to discharge against a head, when, of course, a check valve should be fitted on the delivery side.

Of traps coming under this class the following may be mentioned and afterwards described as being typical

FIG. 24.—Lancaster Steam Trap: Old Type—Section.

examples, each having its own characteristic features and points of material interest:—

FLOAT.
- (1) The "Lancaster" Trap.
- (2) The "Nightingale" Trap.
- (3) The "Syphonia" Trap.
- (4) Wright's "Emergency" Trap.
- (5) Wright's "Victor" Trap.
- (6) The Williams Steam-operated Trap.

THE LANCASTER TRAP.

The immense sale which this trap enjoys is a guarantee in itself of its general excellence both of design and principle, for it is perhaps the most popular float trap on the market. It is made in a variety of forms to suit individual cases. For instance, in figs. 24 and 25 it is seen with a spherical ball float, whereas it is also now made with a long cylindrical float (figs. 26 and 27) of much the same volume, with a smaller box, but possessing special advantages under high pressures. They are also made with feet projecting from the side or bottom, which make them handy for marine work, for they may be bolted to the deck or bulkhead, as desired.

The design is extremely simple, for it consists mainly of a float E, having a small hole at the bottom, and a

FIG. 25.—Lancaster Steam Trap: Old Type—Plan with Cover Removed.

pipe passing down its centre, this pipe having a deflecting bell N on the top for deflecting the discharge, so that the cover may be removed and the working parts examined whilst in use. The copper ball or cylinder E is connected by an elbow pipe to a hollow quick-threaded screwed spindle, having at its end a valve, the latter being opened and closed by the mere rising and falling of the float.

The action is equally simple. For the purpose of description, which will apply to both types, we will assume that the ball has fallen to the bottom of the trap. The valve is then wide open. All the water that may have collected in the system being drained flows through the hollow spindle and elbow pipe into the ball float, up the pipe, when it impinges upon the deflecting bell N, and away to the outlet. Provided there is an accumulation of water present, the ball remains at the bottom of the trap until steam appears, when the water in the elbow pipe and float is swept out, steam taking its place. Now, the steam being so much lighter than the ejected water, the float is rendered buoyant, and immediately rises to the surface, and in so doing rotates the hollow quick-threaded spindle, and thereby closes the valve. The steam imprisoned in the float rapidly condenses, allowing water

FIG. 26—Lancaster Steam Trap: 1903 Type—Sectional End View.

to flow in through the hole supplied for the purpose, thus destroying its buoyancy and causing it to again sink. If in the meantime any water of condensation has formed and is ready to be drained away, it immediately flows into the float, and the same cycle of operations is repeated. However, if there is no water present, that which has already gained admittance to the float through the small hole is swept out through the centre pipe, and, as before, the float again becomes buoyant, floats, sinks, etc. This cycle is being continually carried on, so that it is almost a physical impossibility for the moving parts to "hang up."

The makers have catered for almost every variety of pressure and use, as will be seen from the following classification:—

Class 1.—Is adapted for all cases where large quantities of water have to be dealt with at low pressures, say from 2 to 20 lb. Suited for drying cylinders, bleaching and dyeing keirs, soap-boiling vats, paper-machine cylinders, brewers' pans, etc.

Class 2.—For general draining purposes and steam pressures from 20 to 90 lb. Suited for low-pressure steam mains.

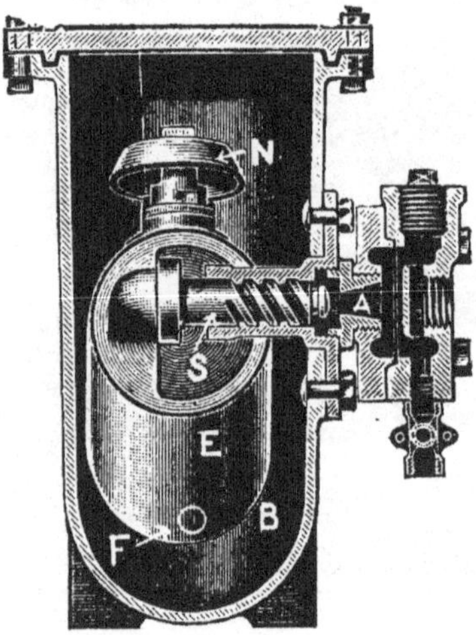

FIG. 27.—Lancaster Steam Trap: 1903 Type—End Elevation.

Class 3.—For higher pressures, 90 to 300 lb. Suited for marine or electric station use.

Class 4.—Specially adapted for lifting purposes.

Class 5.—Marine traps, having projecting toes for bolting either to the deck or bulkhead.

Class 6.—Suitable for high lifts, being made with a much heavier box.

The Nightingale Steam Trap.

Another steam trap working upon exactly the same principle as the Lancaster is the Nightingale trap, a sectional view of one being shown in fig. 28. Their principles of working are identical, the only difference being in the valve opening and closing device, which, as will be remembered, in the Lancaster trap was effected by means of a quick-threaded screwed spindle; whereas in the Nightingale trap the same motion is imparted to the valve by means of two toggle links of manganese bronze changing their angular positions by means of a ball float rising and falling in the box. At the end of the main

FIG. 28.—The Nightingale Steam Trap.

valve stem there is an adjustable air valve, which can be set to open at any desired position of the ball float. The circular fitting in the top right-hand corner is a receiver for the separation of all solid and other injurious matter likely to clog or otherwise prevent the valve from working properly.

THE SYPHONIA STEAM TRAP.

This trap, as its name implies, depends for its working upon a syphoning action, and is in itself a radical departure from other traps of this type. Its somewhat extended use and large sale need no explanation, as it deservedly ranks high amongst its rivals as regards simplicity, reliability, and low cost of upkeep, there being only one moving part, the valve E. Fig. 29 shows a section through the centre. It consists of the usual rectangular box A and a float F, the latter controlling the movement of the valve E. On the left-hand side is a syphon tube H and an air release valve L. The box A having been filled with water and the cover B screwed down by means of the winged screws C C, the float F will rise in the box and lift the valve E off its seat. Upon turning on steam, any water either ready to leave the pipes or that formed by the sudden priming action will flow through the valve until steam

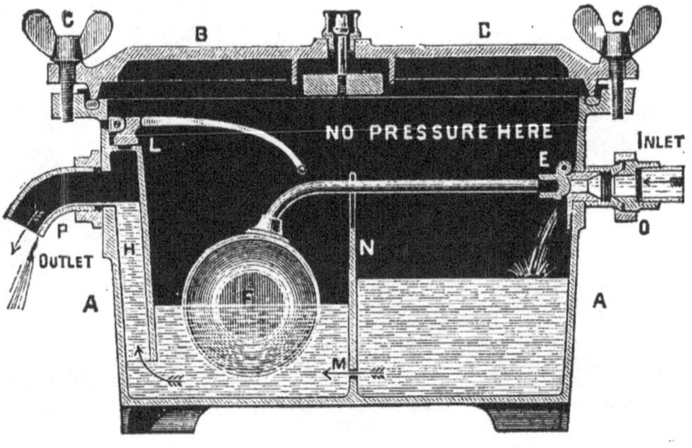

FIG. 29.—The Syphonia Steam Trap—Section through Centre of Trap.

appears, when the water in the body of the trap will rush away to the outlet, through the syphon tube H. Until the valve is above the water level the pressure in the box will accumulate and still continue forcing the water out until the valve is so far closed that the natural condensation of the box is in what may be termed equilibrium with the amount of steam passing, so that, virtually speaking, the valve is never entirely closed; and the cost of expelling the water is represented by the continual condensation of steam taking place, which is of itself necessary for the proper working of the trap. It must not be assumed that this condensation represents anything considerable, for there is never more than a very slight pressure under the cover. A blow through may be obtained by removing the cover and raising the float and lever F. L is an automatic air lever actuated by the float, to discharge the air on starting. The manufacturers of this trap, fully realising that it is practically impossible to construct or design a float trap to work equally well under any condition or to satisfy every specification, have arranged their traps

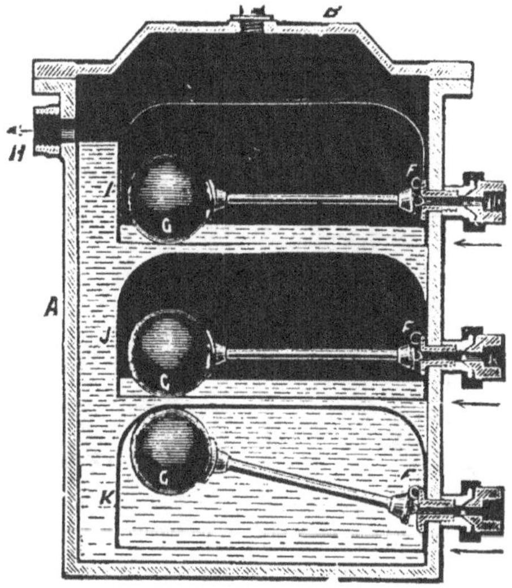

FIG. 30.—Three Modified Syphonia Traps placed in one Box.

into six classes, varying as regards pressure, lifts, etc. The most interesting of these six classes is that known as Class F, which are especially convenient where a large number of traps have to be placed in a relatively small space. A single trap consists of any number—in this case three—of modified Syphonia traps, as shown in fig. 30, I, J, and K (inside a common box B) all having separate inlets, but a common outlet H. The hoods are of copper, and are open at the bottom. Whilst water is passing in any quantity through any or all of the valves—as, for instance, in the bottom trap K, in fig. 30—the float G is at the top of the copper hood until steam appears, when the water is swept out, steam taking its place, so that the float will fall and partly close the valve, as seen in I and J, the action being almost identical to that of a simple Syphonia. It is quite immaterial as to how the pressures may vary at the respective inlets, for they may be, say, 20 lb., 40 lb., and 80 lb. respectively without affecting each other. As will be readily recognised, the system has many distinctive advantages.

The working parts of a Syphonia trap, especially the valve, are easily and quickly inspected or renewed; and last, but not least, the trap is not subjected to any strain likely to cause leakage or broken parts due to sudden

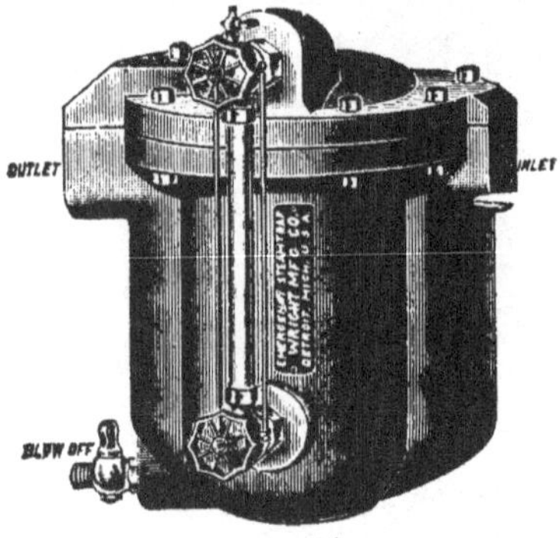

FIG. 31.—The Wright "Emergency" Trap.

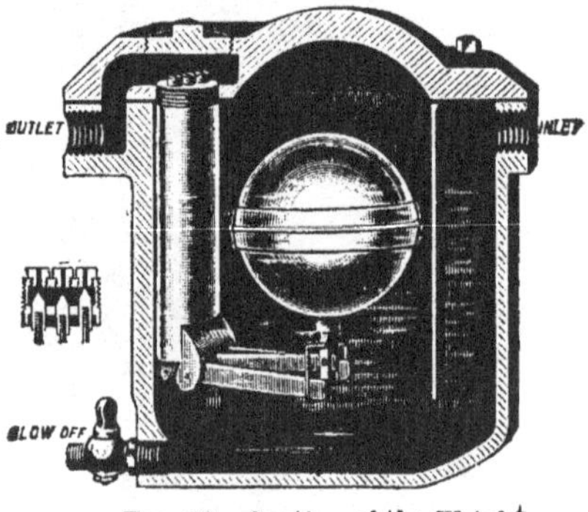

FIG. 32.—Section of the Wright "Emergency" Trap.

unexpected rises in pressure above the normal. These remarks apply equally well to all of the traps of the float type.

MESSRS. WRIGHT'S EMERGENCY TRAP.

The Wright Emergency trap, manufactured by the Wright Manufacturing Company, of Detroit, U.S.A., is one of the most popular types of float-operated traps used in the States. Attention is drawn to fig. 31, which shows an outside view of the trap, in which the most striking feature is undoubtedly the massiveness of the body, combined with the fact that all connections, etc., are most accessible and of ample strength. A cross-section through the working parts is shown in fig. 32, in which it will be seen that extreme simplicity and freedom from small parts is quite a marked feature. It is essentially a type of trap which could be employed with advantage in cases where the amount of water to be drained away is of a very fluctuating nature, as, for instance, in the large steam laundry machines now so much used. It is, indeed, a novel departure from accepted designs; the most interesting point is undoubtedly the unique valve arrangement, which automatically adjusts the area of its discharge to

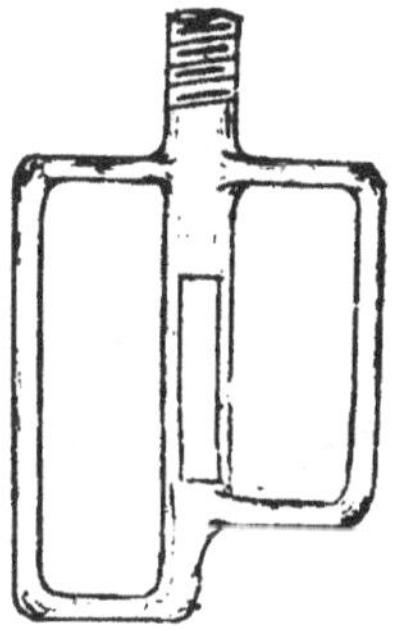

FIG. 33.—Valve Stirrup of "Emergency" Trap.

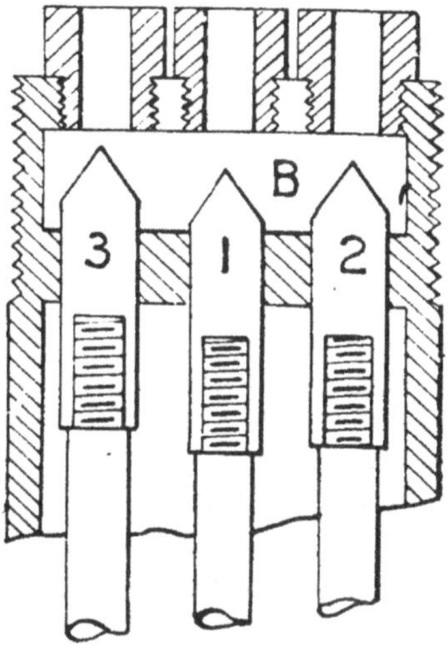

FIG. 34.—Treble Valves of the "Emergency" Trap.

suit the demand made upon the capacity of the trap; for it has three complete valves, situated at the upper end of the cylinder, seen to the left of the float; this, incidentally, is the best possible valve position, owing to it being farthest removed from the places where sediment is at all likely to accumulate.

The action of the trap is as follows: Water entering at the inlet at the right-hand side is deflected by the baffle plate to the bottom of the trap box, and when it has about filled it to the height shown the copper ball floats. The latter has at its lowest point a treble-slotted stirrup, shown separately in fig. 33, which engages with three levers in succession, and operates the valves in a manner which I will now describe.

At the upper end of the cylinder there are situated three entirely distinct discharge valves; these I have shown more clearly in fig. 34, which is an enlarged view of a section through them. Valves 1, 2, and 3 (fig. 34) being connected to their respective levers, it is obvious that the first action of the float is to raise the central lever, which, as seen in fig. 32, is directly connected to the stirrup in the central slot, thereby opening valve No. 1, and allowing any further accumulation of water to drain away. Should this valve opening be of insufficient area to drain away the condensation as fast as it enters the trap, the float will

FIG. 35.

Wright's "Victor" Trap.

FIG. 36.

Section through the "Victor" Trap.

still continue to rise, and will then engage the second lever with the right-hand slot in the stirrup, which, in turn, will open valve No. 2. Should, however, the water be coming in such quantities as to find the combined areas of valves 1 and 2 still insufficient to meet the demand occasioned by such a sudden flush, the ball float will rise still further, bringing the third lever into engagement with its left-hand slot in the stirrup, and thereby opening the third valve, No. 3. In this position all three valves are held open, giving a discharge area about equal to that of the inlet. It is very seldom, however, that such a heavy demand would be made upon the capacity of a trap; but should such occasion occur in practice, this trap would be found quite competent to deal with it effectually.

An additional feature, which is well worthy of comment, is the provision of the baffle plate, which compels the water of condensation to enter at the bottom of the trap and pass upward to the outlet; for when the flow decreases, the float will slowly sink to about half way down the trap; and when in this position the valves would be all closed, and will be protected by a double water seal, for the lower end of both the cylinder and the baffle will be at all times below the water level

A constructional point, which somewhat ensures faultless operation of the valves, is the provision of the bar B (see fig. 34), which holds the three valves in position, and effectually prevents them from becoming decentralised. Further provision has been made for easy access to the valves and other moving parts by so constructing the trap that by removing a few nuts the cover, to which the whole of the mechanism is attached, may be removed bodily; a fact which is much appreciated where adjustment is found necessary.

The Victor Low-pressure Trap.

The Wright Manufacturing Company also make an excellent low-pressure trap for working under any pressure up to 20 lb. per square inch. An outside view of it is shown in fig. 35, but a better idea of its principle will be obtained by a study of fig. 36, which represents a cross-sectional view. A point which will again be evident at first sight is the extreme simplicity and small number of working parts. Its method of operation may well be described as being on a par with its appearance for simplicity, for it is of so simple a nature as to scarcely require a description. It is provided with a very large valve, which, on account of the multiplying action of the float lever, cannot possibly hang up, neither can it clog, for by virtue of its position in the top of the body it is extremely well protected from sediment or any other foreign matter; it is also provided with a double water seal, similar to that in the Emergency trap. From the nature of its design it is an exceedingly cheap trap, and is admirably adapted for use under conditions entailing a somewhat heavy condensation at a low pressure.

The Williams Steam-operated Trap.

Quite an ingenious type of float trap is that manufactured by the Williams Gauge Column Co., of Pittsburg, U.S.A., a part section of which is shown in fig. 37. It is the only trap familiar to the writer which takes advantage of the leverage effort exerted by a float in such a manner as to cause it to operate a valve, permitting steam pressure to act upon a diaphragm, the movement of which, due to its flexibility, in turn opens the main discharge valve.

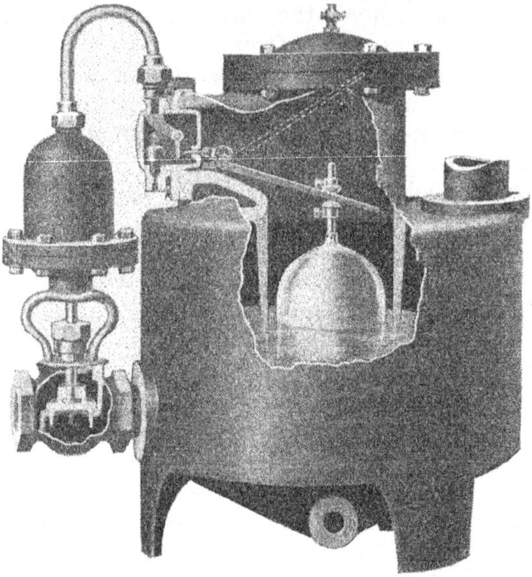

Fig. 37.—Part Section of the Williams Steam-operated Trap.

Dealing with the design in detail, the body or box consists of a strong cast-iron chamber, each part being of more than ample strength and rigidity for the purpose in view. As will be seen, the upper half of this casting is of about half the diameter of the lower portion, and is continued downward towards the bottom, thereby forming a cylindrical-shaped compartment, the bottom or inner half of which serves to maintain the float in position and acts as a guide for it, whilst the upper half contains the valve-operating levers. In the left-hand top corner will be noticed a separate little box, containing two small valves, bolted to a flange on the side of the body, and having a thimble connection at the top leading to a dome-shaped casting. This dome-shaped casting contains a specially-constructed flexible diaphragm, which, when acted upon by the steam pressure in the system under drainage, depresses the valve and holds it hard on to its seating, thereby preventing the escape of the contents of the trap. But when this flexible diaphragm is relieved of the steam pressure, the valve is released and opens, under the influence of the pressure acting upon its under face. It is now perfectly evident from the above that, in order to obtain a controlled discharge, some means must be adopted by which the up-and-down motion of the float alternately releases and admits steam pressure into this diaphragm chamber, so as to open or close the discharge valve. This is obtained in the following manner: When the float is in the position shown in fig. 37 the discharge valve is held "hard on"—it being in the filling position—but water of condensation, entering at the connection on the right-hand top side, causes the float to rise until the lever is in the position shown dotted, when the little valve at the opposite end of this lever is closed, this having the effect of cutting off the steam pressure from the diaphragm chamber; any remaining pressure is further relieved by the small relief valve, which opens to the atmosphere. The latter is operated by the stem of the first little valve acting upon the suspended bell-crank lever. Now, the pressure in the diaphragm chamber being relieved, the discharge valve is opened and a discharge takes place, when the water level sinks and the pressure and discharge valves are again operated in the reverse manner. It will, no doubt, be observed that the main discharge valve is well protected by the usual water seal, and that it is of very liberal dimensions, well calculated to deal effectually with a sudden flush.

One or two constructional features which cannot fail to be of interest are embodied in the design. The float is manufactured by a unique and ingenious process, by which it is rendered extremely strong without impairing its lightness. It is constructed of sheet steel, with a covering of sheet copper, and is placed in an electrolytic bath, where a covering of copper is deposited upon it. A reinforcing band of laminated steel is bound round the centre, which gives it great strength and renders it remarkably free from any tendency to collapse under high pressures. After this it is again placed in the electrolytic bath, this latter process being calculated to prevent any possible chance of leakage and also to render the metal less likely to suffer from corrosive action. In order to test the efficacy of this treatment, and to emphasise the advantage of its employment, the float is tested under a hydraulic pressure of about 750 lb. to the square inch, and is additionally subjected to a temperature of about 300 deg. Fah., in order to discover any imperfections or flaws that may possibly exist in the metal. One can well understand that, after passing such rigorous tests, the makers receive very few complaints about their floats failing in use; for they are guaranteed to work under any pressure up to 500 lb. per square inch.

The diaphragm is also the subject of a special process, for it is constructed of six layers of No. 10 sail cloth between alternate layers of Para rubber, the outer faces being of the sail cloth. This type of diaphragm is possessed of very great tensile strength, and, when properly constructed, is far superior in some respects to a sheet-brass disc. The discharge valve and its seating are of hardened nickel, and are renewable.

CHAPTER IV.

CLASS 2 (*continued*).—BUCKET-TYPE STEAM TRAPS.

THERE are two steam traps of exclusively British design and manufacture which we will now describe. They are (1) the "Sentinel" steam trap, and (2) Messrs. Holden and Brooke's new bucket trap.

The other traps of this class mentioned in the above list, although classed with the three already described, are, paradoxically speaking, of an entirely different type. Their principle is very simple, being nothing more or less than a cylindrical bucket floating and sinking, and in so doing closing and opening a valve, which is fixed to the bucket itself. They are specially adapted for high pressures, and their discharge is, in the opinion of the writer, the nearest approach to perfection so far obtained; for it is an impossibility for anything but water to pass the valve. For the reason that as soon as the water level in the bucket has so far fallen as to render it buoyant, it (the bucket) commences to rise, and closes the valve with a snap, so that the water level in the bucket can never fall low enough for live steam to pass the valve. Due to this property the discharge is very positive, as it never takes place until the bucket sinks, which it can only do when full; so that at each discharge almost a complete bucketful of water is ejected. The moving parts are reduced to a minimum—namely, only one—the bucket, of which the valve is an integral part.

THE "SENTINEL" STEAM TRAP.

Fig. 38 shows a sectional elevation of this trap, which, from its strong construction, simplicity of design, and its few working parts, is hard to find an equal. It consists of a substantially-constructed box containing a heavy spun copper float, the only moving part. From the middle of the bottom of this float projects a spindle guided in a guide tube by vanes, and having at its top end a conical valve. By the nature of its suspension—flotation—this copper bucket is absolutely free to rise, fall, or rotate.

The action of the trap is as follows: Water entering at the inlet raises the bucket and holds the valve on to its seat. When the incoming water has reached the level of the top of the bucket it overflows into and commences to fill it. The bucket, when nearly full, loses its buoyancy and sinks, at the same time filling up, and in doing so drags the valve off its seat; the trap then commences to discharge. As soon as the water level in the bucket has fallen sufficiently to restore its buoyancy it (the bucket) rises and closes the valve. The steam pressure on the back of the valve being relieved directly the bucket commences to sink, due to the weight of the bucket and its contents, the submersion is very complete, for it sinks right down to the bottom of the trap, dragging the valve clear off its seating; whereas when the bucket rises, the steam pressure urges the valve up, and closes it with a snap. The system cannot possibly balance and cause the discharge to degenerate into a dribble, since the valve is either wide open or shut tight, its motion being a perfect "stop" and "start" action. Figs. 2 and 3 show this trap previous to and during discharge whilst working at 180 lb. pressure.

During part of the time occupied in discharging the bucket is entirely unsupported, so that it is perfectly free to rotate with a minimum of friction. The water flowing to the outlet through the vertical guide tube impinges on the vanes, which are part of the spindle, thereby causing the whole bucket to rotate. This rotary motion is a maximum just as the valve is closing, so that the energy stored up in the rotating mass causes the valve to re-seat itself, with a grinding action each time discharge takes place. An additional advantage of the vanes is that they keep the guide tube free from sediment, so that it is impossible for the valve to clog, for, besides being placed as far away as possible from places where grit (which is always present) is likely to accumulate, it is additionally protected by the vanes.

This trap has been designed with a view to keeping pace

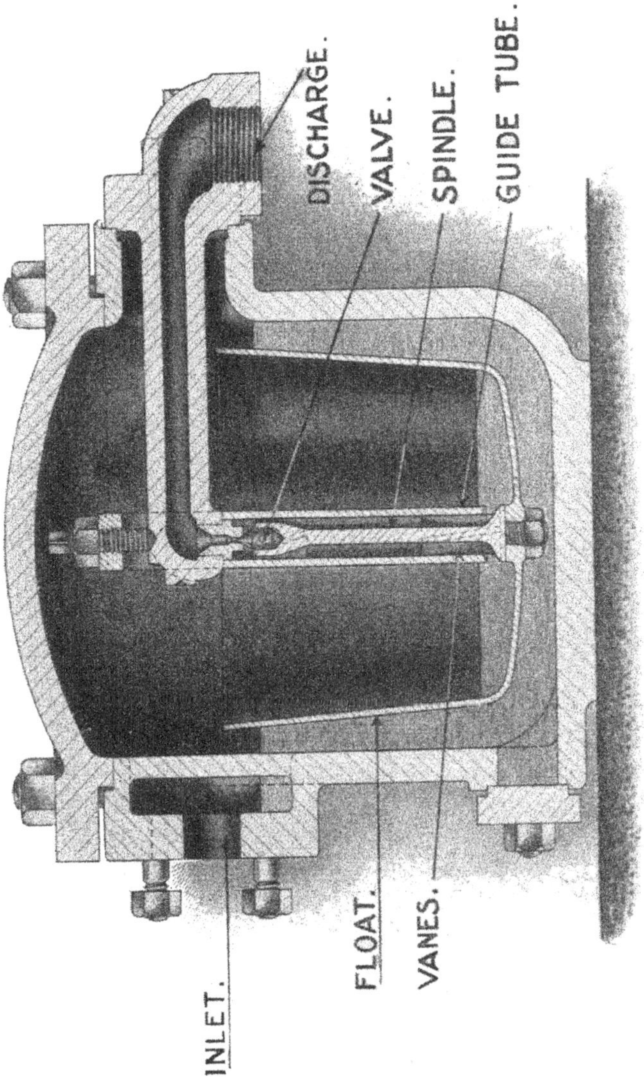

Fig. 38.—Section of the "Sentinel" Steam Trap.

with the modern tendency in the direction of increased pressures, and for this class of work it may be safely recommended, there being absolutely no fear whatever of the bucket collapsing or the box bursting, as they are subjected to a very severe test before leaving the makers, namely, a 200 lb. steam and a 750 lb. hydraulic test.

Messrs. Holden and Brooke's New Bucket Trap.

The other trap of this type, which is also designed for high-pressure use, is the new "H and B" trap shown in figs. 39 and 40. It has only just made its initial bow to the public. It embodies several novel and interesting improvements upon existing traps of this category, chief

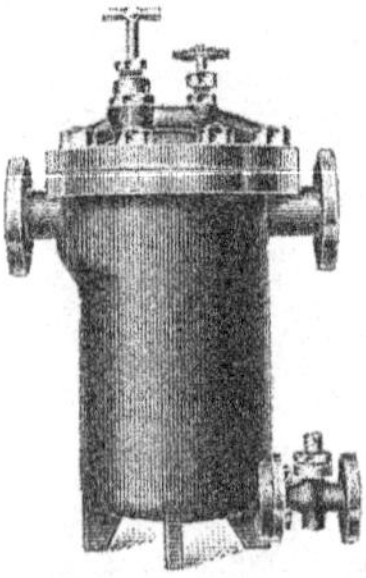

Fig. 39.—The Holden and Brooke Bucket Trap.

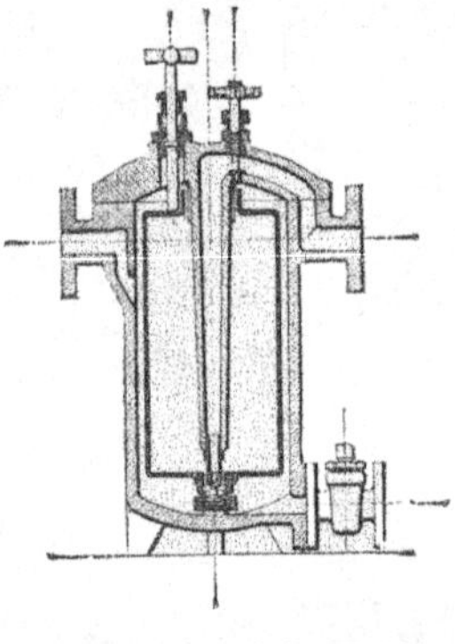

Fig. 40.—Section of the Holden and Brooke Bucket Trap.

among which is the shape of the bucket, which is spun with a bottle neck. Commonly, with bucket traps of the open-topped bucket type, the displacement of the water in the body when the bucket falls is so great as to cause considerable disturbance and vibration of the bucket, thus to some extent preventing an absolutely steady discharge through the valve. However, with a bottle-necked bucket there is only the bottle-neck part above the water line at the time of immersion, and consequently there is only a small displacement in disturbance when the bucket moves. This has a tendency to produce a positive steady action. It is also fitted with an air valve and a blow-through valve, as seen on the top. On the whole this trap is a great improvement in the right direction, and it may be recommended with every confidence, especially for high-pressure use.

Another trap which depends for its action upon a float is Pratt's patent return trap, but, since it is designed for use with a special return-feed system, the writer has purposely left a description over for insertion when dealing with Class 6: Return-feed systems.

The Nason and Sidelug Steam Traps.

The Nason bucket trap has many interesting constructional features, and is shown in fig. 41. It consists of an

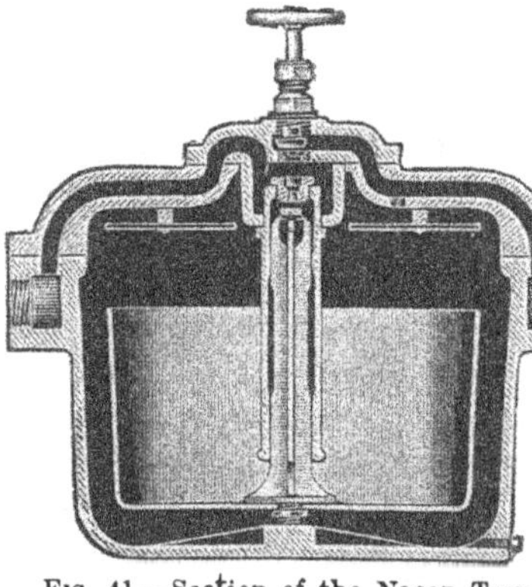

Fig. 41.—Section of the Nason Trap.

open bucket, having a centrally-supported discharge valve, which is actuated in almost the same manner as that in the Sentinel trap. It is the positions of the inlet and outlet passages which are of interest, for, as will be seen, they are contained in the built-up cover. Water of condensation enters at the right-hand side, and, passing through the cored passages, finds its way into the body of the trap through the hole shown, where it impinges upon a diaphragm plate, and is thereby deflected into the space between the bucket and the pot, where, gradually rising, it raises the float and closes the flat-faced valve at the end of the central spindle, the latter being held in place and guided by the housing or sleeve suspended from the cover. A further accumulation of water overflows into the bucket, until its buoyancy is sufficiently destroyed to overcome the tendency of the valve to remain closed, when a discharge takes place, the water in the bucket being forced up the central housing or sleeve by the pressure from the mains acting upon its surface.

When the float is nearly emptied it becomes so light that it is again rendered buoyant, and, rising, closes the valve. The weight of the bucket is such that a water seal always remains over the discharge, which, as will be remembered, is a valuable feature, for it prevents the live steam from reaching the valve and escaping to waste, which, in passing the valve, would have a most deleterious effect upon its surface and seating.

This trap is made in three groups, namely, those for pressures between 1 lb. and 30 lb. per square inch, for use in connection with steam-heating systems and the like; those for pressures up to 70 lb.; and it is also made for higher pressures, in which case the cover is of a much stronger construction. This type is known as the Sidelug.

A hand wheel will be seen in fig. 41, situated on the top of the cover, this being supplied for the purpose of a handy blow through, for by unscrewing it a direct connection to the outlet is opened up. This fitting is also useful when starting up, for, in common with other bucket-type traps, the pot is liable to become air-bound after standing.

The Greenaway Steam Trap.

This trap is shown in section in fig. 42, and, as will be seen, its general arrangement is practically identical with that of the other bucket-type traps already dealt with. However, the discharge valve is of very original design, and is like nothing else of its sort. We do not propose to deal with this trap at all fully, but cannot pass it by without drawing the reader's attention to this valve arrangement. It consists of a piston-shaped valve sliding freely in a cylinder of slightly larger internal diameter. The piston has a small hole through its centre, this latter being normally closed by a small needle valve actuated by the bucket. Now, when the body of the trap is about half full of water the bucket will be floated up so that this little needle valve will close the small hole, or "pilot" valve, as it is called, and hold the discharge valve up on to its seating. But when the water of condensation has risen and overflown into the bucket, the latter, in sinking, opens this pilot valve, which immediately submits the discharge valve to full steam pressure on both faces, instead of only on the lower or under faces, as previously existed. This renders the piston-shaped discharge valve out of equilibrium, for gravity tends to pull it down, and it is relied upon to do so—hence the object of making it an easy sliding

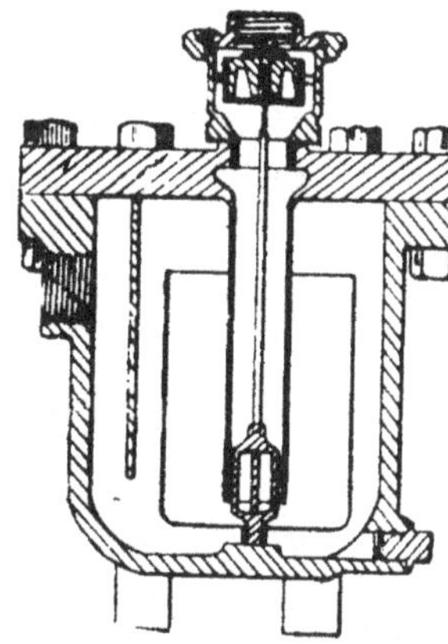

FIG. 42.—Section of the Greenaway Trap.

fit in the cylinder. Although the writer has never seen one of these traps in actual use, he is very fascinated with the design, and, provided he could assure himself that the piston-shaped valve would always remain a good sliding fit in the cylinder, independent of the amount of grit and injurious matter which is always liable to pass, he would have no hesitation whatever in recommending it for trial. One may conclude, in the absence of any information on this point, that the rush of water past the valve would carry this injurious matter out of harm's way, otherwise the trap would not come in for such a large share of patronage as is already extended towards it.

Fig. 43 shows a separate view of the valve, in which its construction, together with the arrangement of the "pilot" valve and its needle, is very clearly shown.

FIG. 43 —Discharge Valve of the Greenaway Trap.

THE KIELEY STEAM TRAP.

Quite a unique type of bucket trap is shown in fig. 44, for, instead of the valve stem being supported by the centre of the bottom of the bucket and the latter being free to rise or rotate freely, as in some of the traps already described, the bucket is hinged at one side to the body, and the valve stem is supported much out of centre of the bottom of the bucket, thereby taking advantage of the leverage effort exerted by the mass of water in the bucket, which, acting about the hinge as a fulcrum, supplies a very much stronger valve-operating effort; this enables a large valve area to be employed, and consequently a trap of greater capacity for a given size is evolved. This trap is also made for low-pressure use,

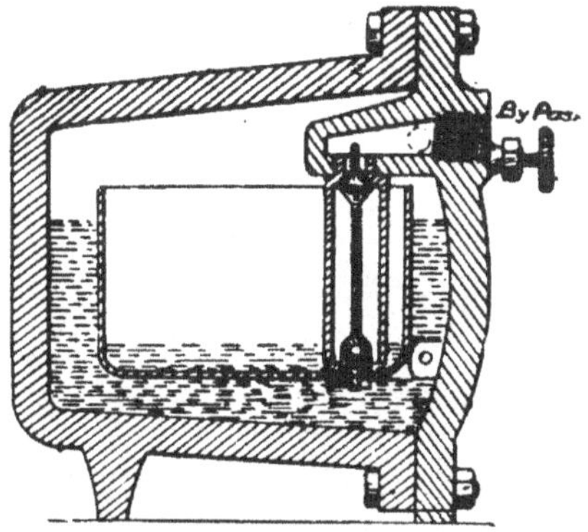

FIG. 44.—Section of the Kieley Trap.

when the valve is much modified, a double-seated balance valve being fitted. These traps, of which the British Steam Specialties Limited hold the British patent rights, are of American origin.

POWER'S STEAM TRAP.

This trap, which is made by the Direct Separator Co., of New York, is another radical departure from orthodox practice in the designs of the majority of bucket-type traps. Shown in section in fig. 45, the trap appears, at first sight, as though it was constructed upon the float principle, owing to the peculiar shape of the bucket, which gives it that appearance; for it consists of a cylindrical vessel, having an inlet funnel at one end and a discharge elbow at the other, this latter being hinged to the body of the trap through the circular sleeve D.

Water entering at the point marked "inlet" is deflected by the baffle immediately opposite into the body of the trap, when it commences to fill the latter, and in so doing floats the bucket, which, riding upon the sleeve D, takes up the position shown dotted in fig. 45. A further influx of water commences to overflow into the funnel, which, accumulating in the bucket, destroys its buoyancy, and as it starts downwards it is flooded and goes to the bottom quickly; in so doing it opens the outlet valve, this being operated by the spindle and collar C. While the valve is open an amount of water equal to about half the capacity of the bucket passes out, the weight of the bucket being so calculated that the water below the level

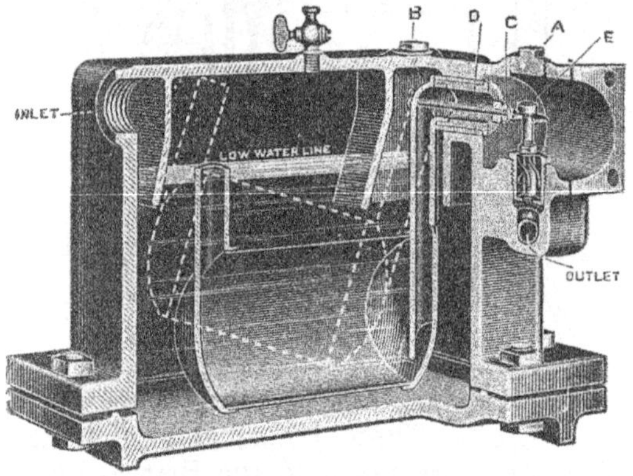

FIG. 45.—The Powers Steam Trap.

of the low-water line floats it up again into the dotted position, and the above operation is repeated. Should the water be coming in any quantity, a continual discharge will take place until the level denoted is reached.

Leaving nearly one-half of the bucket full of water provides a most effective water seal in the form of a U bend. Also the baffles seen cast integral with the top of the box are of sufficient length to prevent steam accumulating in the box. A still further function of these baffles is to diminish the disturbance on the surface of the water when the bucket commences to sink; this, of course, would diminish the tendency of the bucket to sway about, which in time would have an injurious effect upon its point of suspension. Another meritorious feature which is characteristic of this design is the extreme

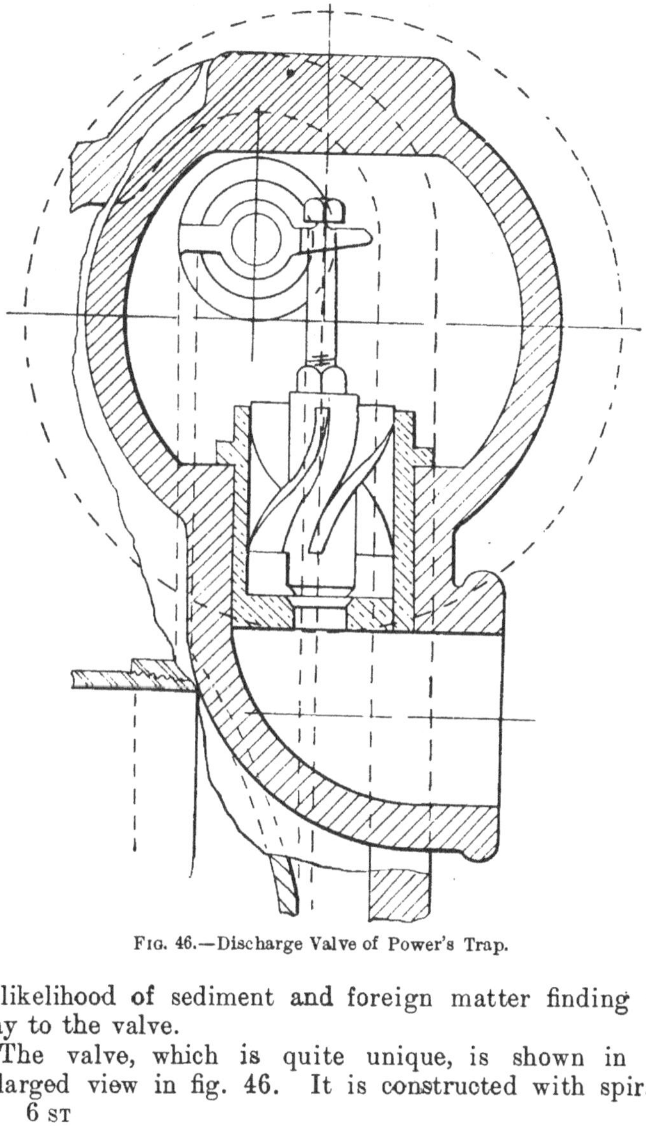

FIG. 46.—Discharge Valve of Power's Trap.

unlikelihood of sediment and foreign matter finding its way to the valve.

The valve, which is quite unique, is shown in an enlarged view in fig. 46. It is constructed with spiral-shaped guides, which, being acted upon by the water as it rushes to the outlet, gives it a rotational motion, which, as already mentioned several times, is a very useful feature, and one conducive to great efficiency.

CHAPTER V.

THERMOSTATIC STEAM TRAPS.

Class 3.—**Those which depend for their working upon the action of a Bourdon tube.**

THE THERMOSTATIC PRINCIPLE.

A Bourdon tube consists of a sealed bent tube enclosing and protecting an expansive fluid; one end is rigidly fixed, whilst the other is free to move. The enclosed fluid is extremely volatile, and is very sensitive to slight variations of temperature. The tube, as will be readily recognised, is the most important, and at the same time the most expensive, portion of the whole trap, so that when in the market for this class of trap the purchaser gets far better value for his money by giving a good price than if he were to buy a cheap foreign-made trap. These traps are very suitable for low pressures, but cannot be recommended for high pressures while there are such a number of essentially high-pressure traps on the market. They possess several valuable features, chief among which are their small size and the independency of their position. The valve remains wide open when cold, and the body is never subjected to any great internal pressure.

The two traps, the "Sirius" and the "Twin-tube," have been selected as being representative types of this class.

THE "SIRIUS" STEAM TRAP.

The "Sirius" trap, manufactured by Messrs. Holden and Brooke, is now so well known as to hardly require a description. The tube, which is of excellent material, is shown removed from the trap in fig. 47. The trap (fig. 48) consists of a cast semi-circular box containing a Bourdon tube, the right hand end of which is fixed, being held in place by the spring and adjusting nut shown; whilst the other end carries a special easily-renewable valve, which is free to move with the expansion of the tube and its contents.

When steam is first turned on, the valve being wide

FIG. 47.—Bourdon Tube of Sirius Steam Trap.

open, any water present flows into the body of the trap through the valve in the left-hand bottom corner until steam appears, when the Bourdon tube immediately warms up and expands, closing the valve suddenly with considerable force. It remains shut until a further supply

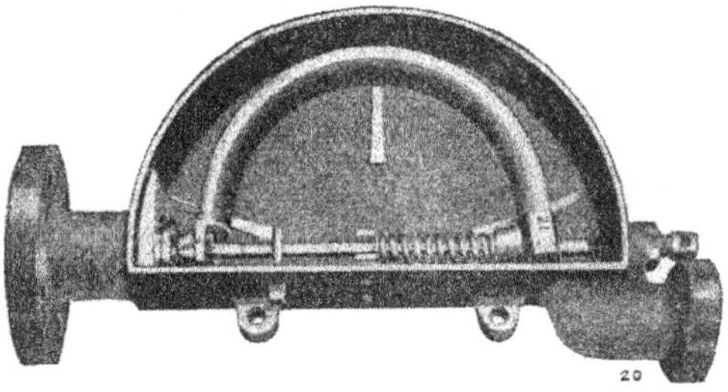

FIG. 48.—Sirius Steam Trap with Cover Removed.

of water cools the tube sufficiently to cause it to contract, when another discharge takes place. The light spring merely retains the tube against the adjusting stop.

THE "TWIN-TUBE" STEAM TRAP.

The other trap of this type, the "Twin-tube," of which Messrs. Smith Bros., of Leicester, are the British agents, is shown in fig. 49.

The working principle of this trap is precisely the same as that of the "Sirius" trap, the only difference being in the shape of the body. It is claimed for this construction that the action of the Bourdon tube, both in opening and

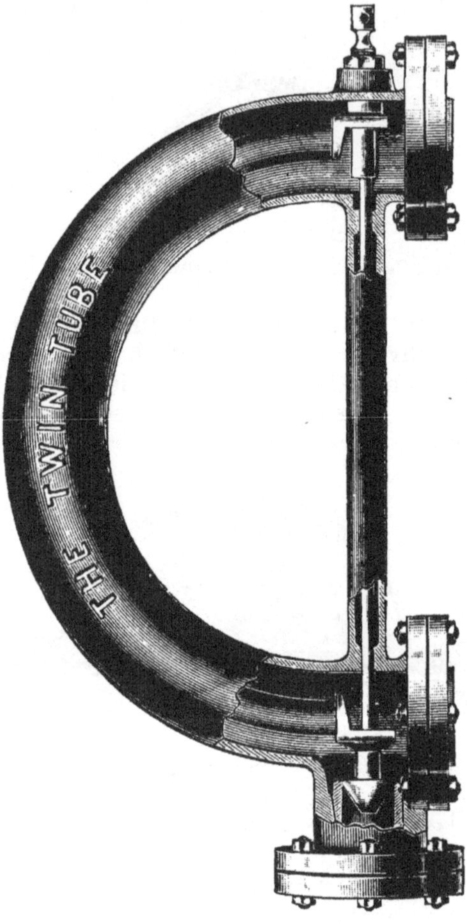

FIG. 49.—Part Sectional Elevation of the "Twin-tube" Trap.

closing the valve, is greatly accelerated. The writer has never had an opportunity of seeing this trap in actual use, so cannot therefore write upon it at any length. However, whilst passing, one cannot fail to see that it embodies improvements in the right direction. For marine work and fairly high-pressure use the outer case is constructed of gun metal.

Class 4.—**The action upon a corrugated metallic chamber containing a volatile fluid.**

There is another type of thermostatic trap which utilises the properties of an easily-volatilised fluid, though in a somewhat different manner. The volatile fluid is contained in an hermetically-sealed flat chamber, having its opposite faces formed of two corrugated discs of reasonably thin metal.

There are several makes of this type, differing only in detail, but embodying exactly the same principles, namely:—

(1) The "Imp" steam trap.
(2) The "Midget" steam trap.
(3) The "Stills" steam trap.
(4) The "Dunham" steam trap.
(5) The "Horne" steam trap.

The action of the corrugated metallic chamber is the same in each, and may be briefly described as follows: The chamber consists of two flat circular corrugated plates bent over and soldered together at their edges. When the joint has been made absolutely air-tight, the chamber is exhausted of air, and a few drops of the volatile fluid inserted. There is now a partial vacuum in the chamber, which has the effect of drawing the centres of these plates inward when they are cold; but immediately live steam strikes either face the fluid rapidly volatilises, and by this means a considerable pressure is generated inside the chamber. This pressure forces the corrugated surfaces outward. This motion is used to operate a valve. As soon as steam is cut off the fluid rapidly condenses, creates a partial vacuum, and by this medium the discs are again drawn inward, allowing either water or steam to pass, as the case may be. The position of the disc in immediate contact with the valve is so adjusted that the pressure generated in the chamber only closes the valve when live steam appears, but allows boiling water to pass, so that the movement of the valve is purely pulsational.

Like the Bourdon tube in a "Sirius" trap, the corrugated metallic chamber constitutes the only part of the trap which is at all likely to give any trouble, so that too much emphasis cannot be put on the fact that the value of a trap of this class is practically determined by that of the chamber.

Of course they are only applicable for uses entailing but moderate pressures, their small size and weight making them admirably suited for draining railway carriage heating apparatus, especially those in which the storage system of steam heating is employed.

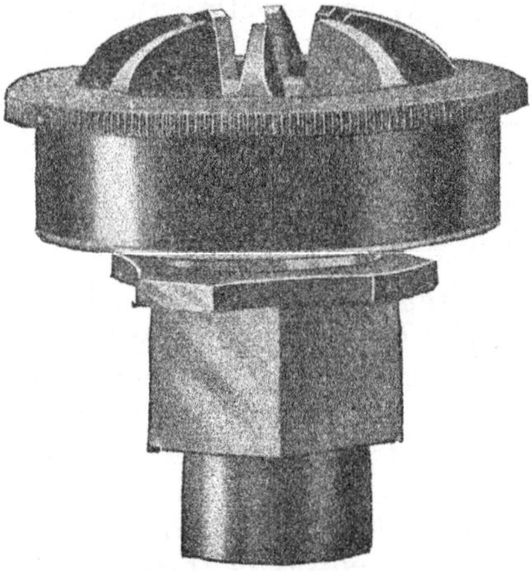

FIG. 50.—Imp Steam Trap.

THE "IMP" STEAM TRAP.

Figs. 50 and 51 show a trap of this class which has been specially designed for use with railway carriage heating apparatus. It is known as the "Imp." A glance at fig. 52—a section of an "Imp" trap—will readily explain its action. Being connected at a suitable point of the system being drained, water of condensation enters at the inlet shown, flows past the valve, round the diaphragm chamber, and out underneath. The position of the diaphragm chamber having been properly adjusted by means of the adjusting ring and lock nut shown, the valve will remain open so long as water is passing, but immediately live steam appears—for reasons previously mentioned—the diaphragm closes the valve sharply, and holds it hard on to its seat. Now, the lower face of the diaphragm chamber being in intimate contact with the surrounding air, as clearly seen in fig. 51, the pressure in the chamber, generated by the volatile fluid, falls rapidly, so that the valve again opens. The writer knows this trap to be exceedingly sensitive, opening sometimes to discharge as little as a teaspoonful of water, the motion of the valve differentiating, so to speak, between the presence of water and that of steam.

FIG. 51.—View from below of Imp Steam Trap.

THE "MIDGET" STEAM TRAP.

Another trap of this class is the "Midget," a section being shown in fig. 53. In this trap the corrugated chamber containing the volatile fluid forms the head, and is also the means by which the relative positions of the valve and the chamber are adjusted, for to set the trap it is only necessary to rotate this head until steam blows through freely, then rotate it in the opposite direction until the blow through ceases, and lock it in this position by means of the winged lock nut supplied.

The action of the chamber is exactly the same as in the "Imp" trap, though it is claimed that, since it is in more intimate contact with the surrounding air, the trap has an increased sensitiveness. This trap is made in a variety of forms; in the pattern "B" type it is fitted with a stop-valve underneath, so that the steam may be temporarily shut off by screwing a conical valve against the lower end of the short tube in the centre of the body of the trap. This pattern is especially adaptable in cases where it is not always convenient to shut steam off from the trap for long; for instance, when the water is dirty and gritty, by closing the underneath stop valve the

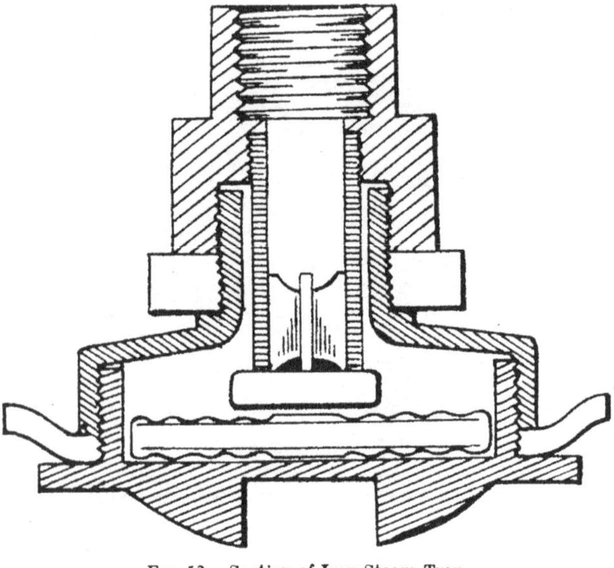

FIG. 52.—Section of Imp Steam Trap.

chamber can be removed and the valve inspected inside thirty seconds. It is also made with a dome-shaped body, which is claimed to have additional advantages for outdoor work.

"STILLS" STEAM TRAP.

Undoubtedly the most interesting and simple adaption of this type of thermostatic valve is that exemplified by "Stills" steam trap. A section is shown in fig. 54 of type A of this make, where it will be seen that one of the most important modifications is in the shape of the diaphragm chamber, for instead of consisting of two flat corrugated plates as in the "Imp" and "Midget" designs, the upper side of the diaphragm chamber is spun with a central cone-shaped projection, whilst the lower side has a similar cone projecting inwards, the latter being of such size as to almost fit inside that formed on the upper diaphragm; so that, as may be seen in fig. 54, these extensions form a conical valve. By this means a much more sensitive action is obtained, as the diaphragm is in intimate contact with the contents of the inlet pipe; for if there is as much as a teaspoonful of water present it must of necessity come into close contact with the cone

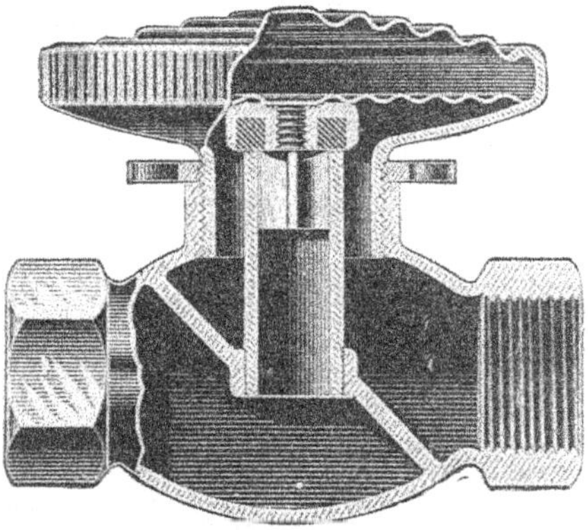

FIG. 53.—Section of Midget Steam Trap.

(no other portion being exposed), so that the cooling action which this water has on the volatile fluid causes the diaphragm to contract sufficiently to admit it to pass.

At first sight it might be thought that from the nature of this conical valve and its knife-edged seating the valve would in a short time become severely scoured; but such is not the case, as it has been proved by experiment that when the trap is properly adjusted the pressure on the knife-edged seating is never much more than 2 lb. above the pressure in the pipe being drained.

The adjusting screw D, which is hollow, is of just sufficient length to close the trap when desired without being able to do more than slightly nip the two conical walls of the diaphragm between its head and its seating. When the trap has been properly adjusted, the adjusting screw D may be locked in position by the lock nut E.

This type of trap, being very small and light, is very convenient for hanging from the end of a small draining

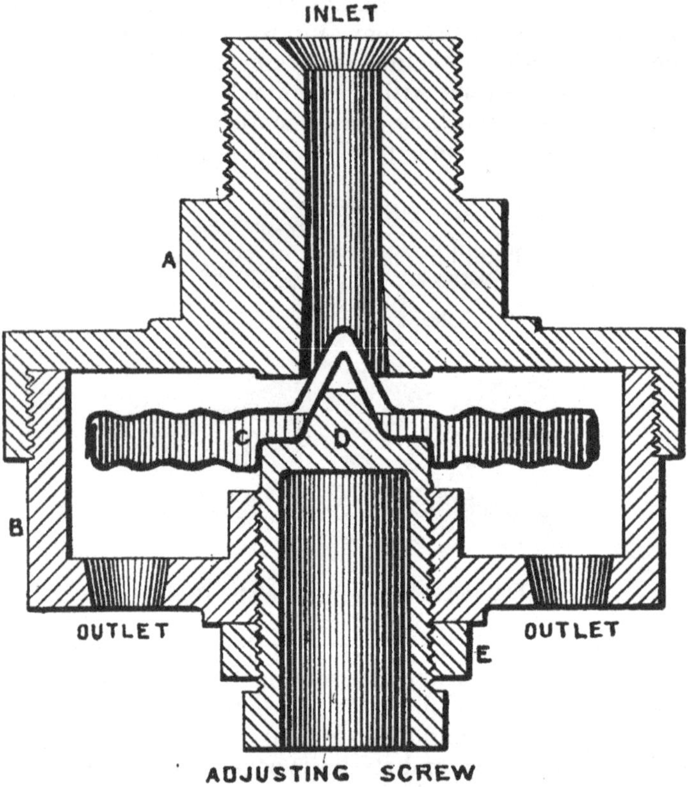

FIG. 54.—The "Stills" Trap.

pipe; and since the lower portion B of the case is easily unscrewed, it is by no means necessary to break a joint in order to inspect or renew the diaphragm chamber.

The characteristic features of this type of trap are embodied in various other designs of the same maker;

the type F being specially adapted for use with steam heaters on coaches, being fitted with independent shut-off and blow-through levers.

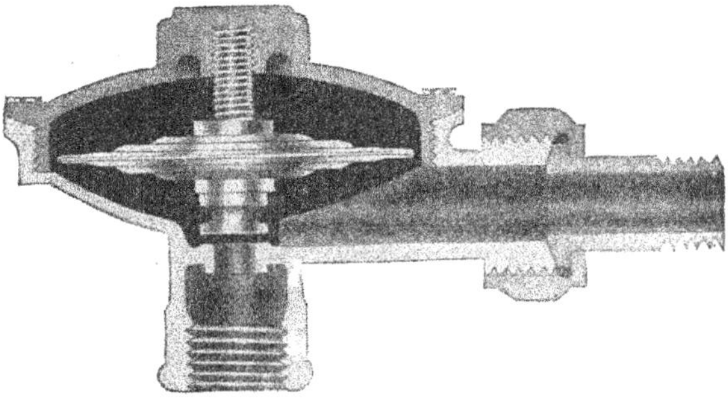

FIG. 55.—The Dunham Steam Trap.

FIG. 56.—The Dunham Trap.

THE DUNHAM STEAM TRAPS.

Messrs. Dunham, of Iowa, U.S.A., also manufacture thermostatic traps very similar to those previously described, some of which are worthy of attention. Fig. 55

shows the type they intend for use in connection with steam-heating systems, it being intended for use upon the return end of radiators when a difference of pressure exists between the radiator and the return pipe. The same force of expansion is used as in other types of thermostatic traps mentioned, and it also possesses many of the same advantages, there being but little actual difference, except, perhaps, in the shape or type of valve.

Fig. 56 shows another, made up in a slightly different manner, this pattern being fitted with a conical valve; also, its inlet and outlet connections being in the same straight line, it may be employed with advantage in some cases.

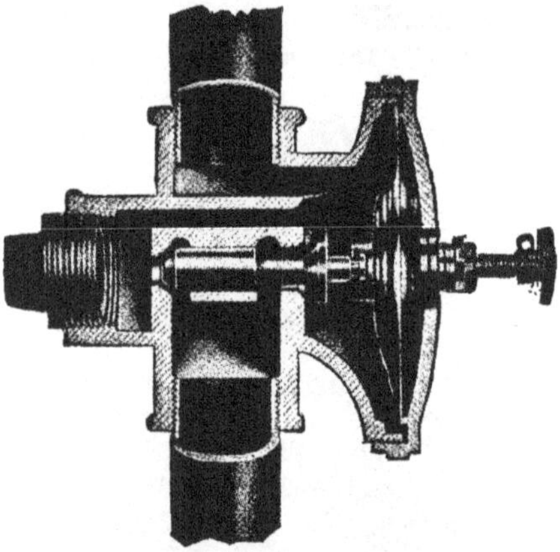

FIG. 57.—The Dunham Trap.

A very neat type is shown in fig. 57, the body being of phosphor-bronze, which is the best steam metal on the market; also the valve is of a very different design, being capable of a very heavy discharge capacity, comparatively speaking.

THE HORNE STEAM TRAP.

Another trap employing the expansive property possessed by a volatile fluid when under the influence of varying temperatures is that shown in section in fig. 58. In this trap the expansive fluid is contained in the hollow tube A, it being sealed at one end, whilst the other is closed by a small piston P. Half way along this tube, and integral with it, is the valve F. When live steam is present the expansive fluid forces the containing tube away from the piston, bringing the faced portion of the

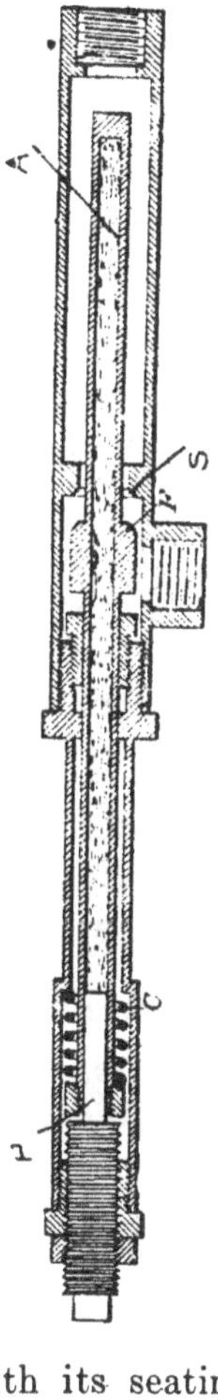

FIG. 58.—Section of the Horne Trap.

valve into contact with its seating S. But when water is present, owing to the tube containing the expansive fluid projecting into the inlet pipe, a very rapid cooling of the fluid takes place. The fluid contracting causes the tube to again slide further on to the little piston, as shown in the illustration, when a discharge takes place, and the operation is repeated.

The arrangement is certainly original, but, as the writer has never had the opportunity of seeing it in actual use, he cannot fully appreciate all its advantages. The fact of relying entirely upon the expansion of the fluid to operate the valve direct, instead of employing a corrugated diaphragm, appears upon the face of it to be an advantage, though as to how far this is borne out in actual practice cannot be dealt with here.

It is also an advantage to employ the expansive fluid in the form of a thin rod instead of in a bulk, as by so doing the net amount of expansion would take place in the direction in which it would be most beneficial, namely, in the length of the rod; thereby obtaining a larger valve opening.

It is claimed by the makers, Messrs. Baird and Tatlock, of Glasgow, that between atmospheric conditions of temperature and that of live steam there is a movement of fully $\frac{1}{2}$ in. at the valve. The method of adjustment is very easily got at; the same may also be used as a blow-through. The over-all length of the trap is 18 in.

CHAPTER VI.

A Differential Water-pressure Steam Trap.

Class 5.—**Those which depend for their working upon the principle of differential water pressure, that is, to the difference of pressure due to two unequal columns of water acting upon the opposite faces of a flat diaphragm, the motion of which controls the valve.**

There is only one trap used at all extensively which may claim to come under this class; it is the "Flinn" trap, the invention of an American engineer, Mr. R. J. Flinn, in whose country it is used to a great extent in connection with almost every branch of steam engineering;

having proved itself very valuable in the draining of the steam chests, etc., of the high-powered pumping engines, which are such a feature of the waterworks of our American cousins.

Fig. 59 shows a sectional view through the working parts of this trap, which consists of two wrought iron pipes of different diameters, the smaller one being placed inside the larger. At the lower end of these pipes is a cast-iron shell, divided into two compartments, F and Y, by a flexible diaphragm D. The contents of the larger or outer pipe communicate with the bottom compartment Y, bounded by the under side of the diaphragm; whilst the smaller or inside pipe communicates directly with the compartment F on the upper side of the diaphragm. In the latter compartment the valve V is clearly seen, being held against the diaphragm by the light spiral spring S, the tension of which may be adjusted by means of the lever L, and the lock nuts on the screwed stud on the outside. The upper end of the external tube supports the inlet chamber C, which is of a globe-shaped appearance, having a peculiar inverted funnel-shaped entrance, by which means the incoming water is distributed to the best advantage for the proper working of the trap.

The action of the trap is as follows: Water of condensation entering at A, follows the surface of the funnel, and is thereby directed to the lower chamber Y, through the annular space between the two tubes, until the latter is completely filled, when it overflows into the chamber F by way of the inner tube. Now, when this inner tube is about half full, the diaphragm D is in a state of equilibrium; the pressure on the bottom or under surface due to the column of water X just balancing that on the top or upper surface, due to the pressure of the valve spring S, together with that of the column of water in the inner tube; under these conditions the valve is on the point of opening. The outer column of water acting on the underside of the diaphragm tends to hold the valve hard on its seat, whilst that in the inner tube, together with the action of the spring, tends to drag it off, so that it is clear that when the two equal and opposite forces exactly counterbalance each other, it is only necessary for the incoming water to add to the head in the inner tube, in order that the valve may be made to leave its seating and allow a discharge to take place, until the head in the inner tube has fallen to a level, approximately

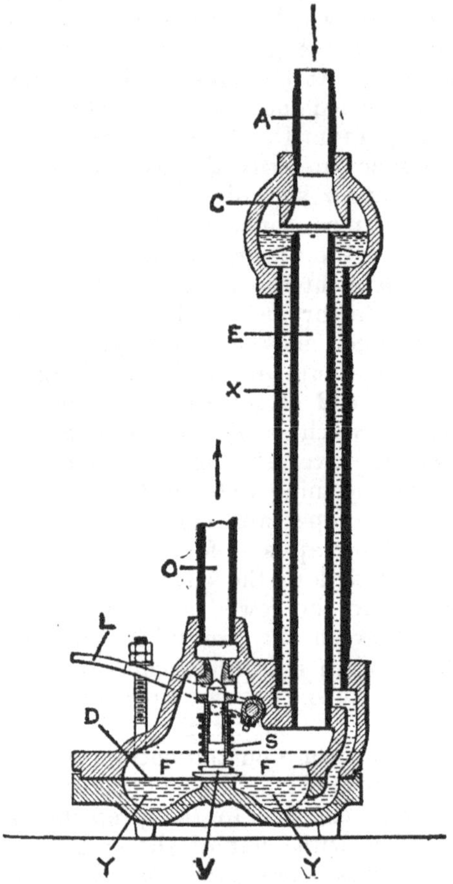

FIG. 59.—The "Flinn" Trap.

a little below the middle of the pipe E. This level may be called the "water seat level," for the valve opens every time the water rises above this level, and closes every time it falls below it. So that the discharge is entirely controlled by the quantity of water needing ejectment—a very important feature.

It will be recognised at once that the great simplicity arrived at in this trap is entirely due to the fact that the controlling forces are in no way whatever dependent upon either the temperature or the pressure of the contents of the system being drained; for the pressure may be anything from 1 lb. to 300 lb., and the trap will work equally well with but one adjustment.

One feature which cannot be passed without special mention is the absolute protection afforded the valve whilst working, for nothing but water can pass it, owing to the fact that before steam can appear the water level must fall to the bottom of the inner tube, which is practically an impossibility as long as the trap is in working order.

The lever L shown in fig. 59 has a twofold function: firstly, it may be employed to adjust the pressure of the spring S, so as to either raise or lower the water seal level; and secondly, it performs the all-important function of a simple and handy blow-through lever.

Accessibility of the valve is, in the opinion of the writer, the only weak point, though as to the direction in which this may be improved it is not easy to say; the fact of steam never passing the valve seating except at a blow-through, it does not necessitate such frequent renewal or inspection of this part, as is essential for the efficient working of some other traps. This trap can be recommended for almost any and every class of work, for it is the only steam trap on the market that embodies such simplicity with an equally wide range of possible adaptions.

CHAPTER VII.

Return-feed Systems.

Class 6.—**Return-feed Systems.**

To anybody taking the trouble to measure the total quantity of water discharged during 24 hours from the steam traps of a fairly large installation, an invention enabling one steam trap to efficiently drain the entire pipe range, and to return the water thus drained directly back into the boilers, together with its large store of sensible heat, will appeal very forcibly, not only from its economical aspect, but equally so from that of its simplicity.

In many instances where one or another of these systems has been adopted the economy thereby entailed has always come out well on the right side, and in many cases a very marked decrease in the annual coal bill has been observed.

There are two return-feed systems familiar to the writer which are well worthy of consideration; they are known as Pratt's patent return steam trap system and Holly's patent gravity return system respectively, and will be hereinafter known as Pratt's and Holly's system, as the case may be. The former of these two systems—namely, Pratt's—is entirely dependent on the action of a specially designed steam trap, and may therefore be described in full, without falling outside the limitations embodied in the title of this series on modern steam traps. Not so, however, regarding Holly's system, for, being entirely dependent upon gravity, as its name implies, and not requiring a steam trap anywhere in the system, it cannot, properly speaking, be entitled to a place here. But after all is said and done, it performs all the functions of an efficient steam trap with the addition of its returning capabilities, and it is so full of interest that the writer has not hesitated to include it here, albeit only a short description.

PRATT'S PATENT RETURN STEAM TRAP SYSTEM.

This system, of which Messrs. Wm. Whiteley and Sons, of Huddersfield, are the sole manufacturers, is the most ingenious and unique adaptation of a steam trap and discharge working upon the return-feed principle so far in vogue. Shown in fig. 60, it is used for returning direct back into the boiler, automatically and continuously, the condensed water which is produced in the various steam-consuming appliances, so that it enters the boiler at a temperature as near as possible to that of the water contained therein.

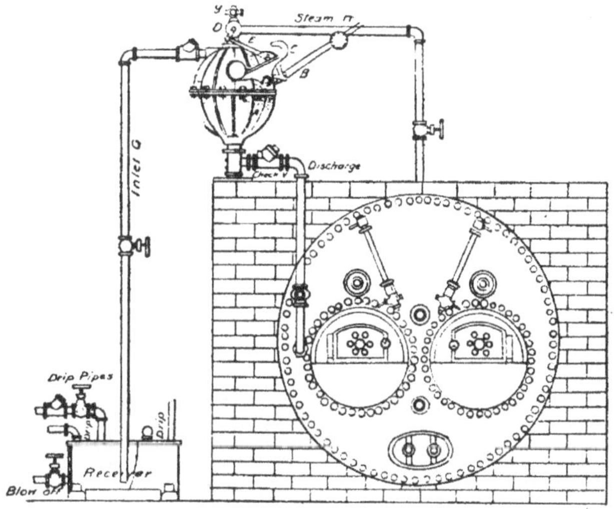

FIG. 60.

Briefly, the system is as follows: All the various drips from their respective sources are led into a cast-iron box known as a receiver, this box being placed below the level of the lowest drip. The water thus accumulated in the receiver is forced or drawn up, as the case may be, into the body of the steam trap, which consists of a float inside an egg-shaped casting, situated at not less than 4 ft. above the water level in the boiler. As this casting fills up, the float rises until a certain level is reached, when, by a system of levers and weights, actuated in the first place by the movement of the float, it automatically admits steam into the top of the trap at full boiler pressure. The inlet pipe being fitted with a check valve, the water cannot flow back into the receiver. Now let us consider what we have: here is a volume of water 4 ft. to 6 ft. above the water level in the boiler; there is full boiler pressure above and below it; therefore the law of gravity comes into force and the water flows rapidly down the outlet pipe to its own level in the boiler. The float falls with it until just before the bottom of the trap is reached; the full boiler pressure is automatically shut off, allowing the water in the receiver to again fill the trap, when the action is repeated.

Describing the system in more detail, and turning now to the trap proper, as seen in fig. 61, it consists of a cast-iron egg-shaped hollow receptacle A, divided horizontally by a flanged joint, and suitably ribbed to withstand the high boiler pressures now in use. Inside, riding on a fork, which is firmly fixed to the same spindle, to which is attached the external lever B and the balance weight W, is a hollow cast-iron float, the whole being adjusted to a state of equilibrium by means of the movable weight W on the lever B, as shown. A short distance along the lever B is riveted a short steel pin R, which engages with the skate C and its rolling weight R W. This skate works upon a separate shaft from that of the lever. Working in a slot S, at the top of the skate, is a short connecting rod M, which is attached by a short crank N to the spindle of the steam valve D. This valve, which is of a very simple construction, fulfils a twofold purpose, dependent upon the position it is in; for in one position—during discharge—it admits the steam from the boiler into the trap, at the same time closing communication to the atmosphere through the pet cock or air valve P, whilst in its other position—during inlet—it shuts off the boiler steam and admits air at atmospheric pressure through P.

The action of the trap is as follows: Assume it to be in the position shown in figs. 60 or 61. The water of condensation in the receiver enters the body of the trap through the inlet G, the displaced air and vapour escaping through the air valve P. As the water rises in the trap the float rises with it, which, as before mentioned, is balanced by the lever B and its adjustable weight W. The stud R after a while engages with the skate C, and slowly pulls it over into a horizontal position. As soon as it tilts down in the opposite direction the rolling weight W

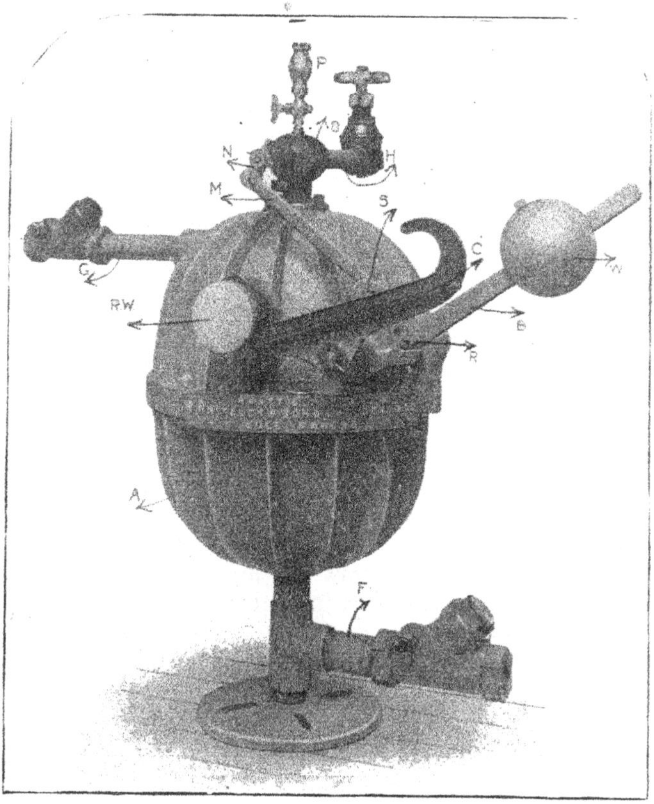

Fig. 61.—The "Pratt" Trap.

runs along the skate and weighs it down suddenly, and in so doing causes the connecting rod M to actuate the crank N and open the valve D, thereby admitting live steam at full boiler pressure on to the water in the trap, which cannot get back along the inlet pipe owing to the automatic check valve placed thereon; so that, as already stated, the water flows by gravity down to its own level in the boiler. The rate of discharge depends solely upon the amount of friction or other resistance due to elbows, bends, tees, etc., in the fall pipe. As the water in the trap falls the float falls with it, until just before the bottom of the trap is reached, when the rolling weight again operates, this time rolling back to its original position, closing the valve D and releasing the pressure in the trap so as to allow another charge of water to enter at G.

The air valve has a special function. Suppose that the water in the receiver is, for a few hours only, a very small quantity, the trap would consequently take a longer time to fill. Now, after it has discharged its contents, the steam remaining when the valve D has closed condenses and forms a vacuum, which is made use of by the air valve P, closing and preventing admission of air. The vacuum thereby formed is in numerous cases often of the greatest importance, as it very materially assists the flow of water from the receiver, which is especially advantageous when the pressure in the receiver is low.

The receiver is a cast-iron steam-tight chamber, which receives the water of condensation in bulk. It is essential that it should be placed below the level of the lowest drip, so that the water may flow into it by gravity if desired. All drips, which must be of the same pressure, and be fitted with separate stop and check valves, are connected to the brass bushes in the top of the receiver. From the centre large bush in the top of the receiver the inlet pipe G runs to the trap, being fitted with a reliable check valve. The water flows from the receiver along this inlet pipe, partly by pressure from the source of the drips and partly by suction due to the vacuum.

In special cases where the condensed hot or boiling water is delivered to the receiver at or near atmospheric pressure, and it is desired to return this to the boiler, special means have to be adopted in order to lift the water into the trap, which, for reasons previously mentioned, must be about 4 ft. to 6 ft. above the water level in the boiler. It is accomplished in the following manner: A second trap of the same type, but for purposes of distinction called a "lifter" trap, is so placed with its necessary stop and check valves that the water from the receiver may flow into it by gravity, when it is forced by steam pressure into the return trap. The steam pressure may be taken from any convenient source, so long as it is amply sufficient to lift from the lower or "lifter" trap to the return trap. The discharge pipe of the lifter trap is connected to the inlet of the trap. The steam pressure forces the water from the lifter trap into the return trap, which in turn returns it to the boiler as before.

The following points should be noticed in connection with this system. The inlet to the boiler should for obvious reasons be placed well below the water level, and the return trap should be placed as near to the boiler as circumstances permit, so that the live steam pressure above the water in the trap should correspond exactly with that existing in the boiler.

One great advantage in these systems is their simplicity, for they do not require skilled attention when in use; also, apart from considerations of economy in water and coal, they very materially increase the life and efficiency of boilers so fitted, for only pure distilled water is returned to the boiler as a continuous feed, thus reducing, and to some extent preventing, the accumulation of scale.

Holly's Gravity Return System.

The other return-feed system cannot, strictly speaking, be considered out of place in this series, for it performs all the functions of a steam trap, without actually including one.

The principle is decidedly novel, but we can only give a short description. Fig. 62 shows the system fitted to a range of boilers, one only being shown. It consists of the following parts: A drip receiver, a main receiver A, two separators I and B, rise and fall pipes C and D, and the necessary connecting pipes. The drip receiver F is a cast-iron vessel to which the various drips are led; from this receiver they flow into the main receiver A through a specially-designed suction tee. The forcing end of this suction tee is connected to the steam separator I, and its condensation is also led into the receiver under full boiler pressure. The water is then swept up the riser pipe in sections of finely-divided particles, due to the special discharge head on one end of

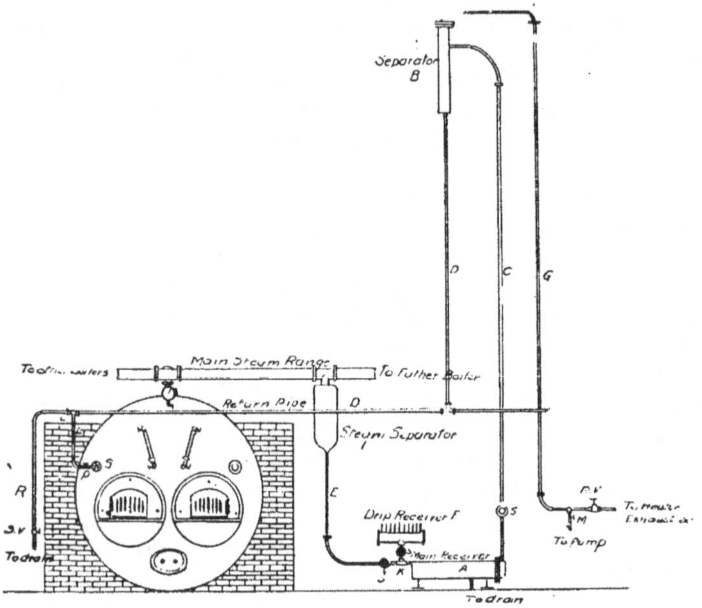

FIG. 62.—The Holly Gravity Return System.

the main receiver into the separator B, when it falls into the down pipe D, and thence direct into the boiler. The water in its upward flight is carried with the steam that is required to maintain proper circulation, and occupies a vertical column in the down pipe above the water level in the boilers, when the weight of this column of water, together with the pressure from the receiver, overcomes the pressure in the boilers, thereby enabling the water of condensation to form practically a continuous feed to the boilers. The object of the small pipe G and the reducing valve R V is to enable the engineer in charge to have absolute control of the circulation.

CHAPTER VIII.

Bundy Steam Traps.

Bundy steam traps are manufactured by the A. A. Griffin Iron Company, of New Jersey, in two different types, those known as return traps and separating traps.

Fig. 63.—The Bundy 100 Series Separating Trap.

Dealing with each type in detail, the return traps are employed specifically to discharge water of condensation into steam boilers against the steam pressure existing therein, although they are very frequently used with success as lifting pumps, to raise water or other liquids from a given level to any higher point, and also as meters to measure the flow of condensation from steam coils, etc.

Attention is drawn to fig. 63, which shows an exterior view of a Bundy 100 series return trap, where it will be seen that it consists essentially of a pear-shaped receiving bowl working on a trunnion at its stem end, and so set to hang suspended that when empty its top surface rests against the top of the circular hoop; but when full, the weight of the contained water, the centre of mass of which is some distance from the trunnion, causes the bowl to fall so that it rests on the bottom of the hoop, remaining in that position until all the water has passed out.

The actual manner in which the discharge is obtained will be understood more readily by a study of fig. 64, as

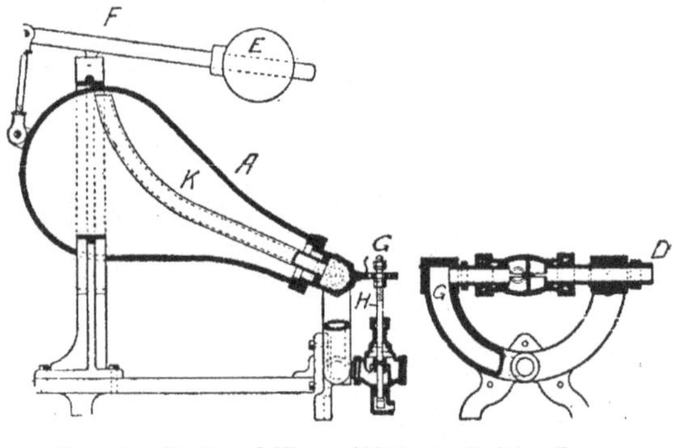

FIG. 64.—Sectional View—100 Series Return Trap.

showing a section through the working parts, the manner in which each operation takes place is more easily followed. The water of condensation entering at the point D, on the right-hand side of the trunnion, passes into the bowl, which, when of sufficient weight to overcome the leverage effect of the weight E, on the horizontal lever F, overbalances and takes up a lower position. Whilst falling the projecting ring G pulls up on the valve stem H and opens the valve I; this being connected direct to the steam space in the boiler admits steam at boiler pressure. The pressure in the trap and boiler now being the same, and the trap being located above the water level of the boiler, the contained water flows into it (the boiler) by gravity. The trap bowl now being empty, the weight E pulls it back into its original position ready for another fill up, and at the same time the ring G closes the steam admission valve and opens the small valve J, the function of the latter being to exhaust the steam and air remaining in the bowl.

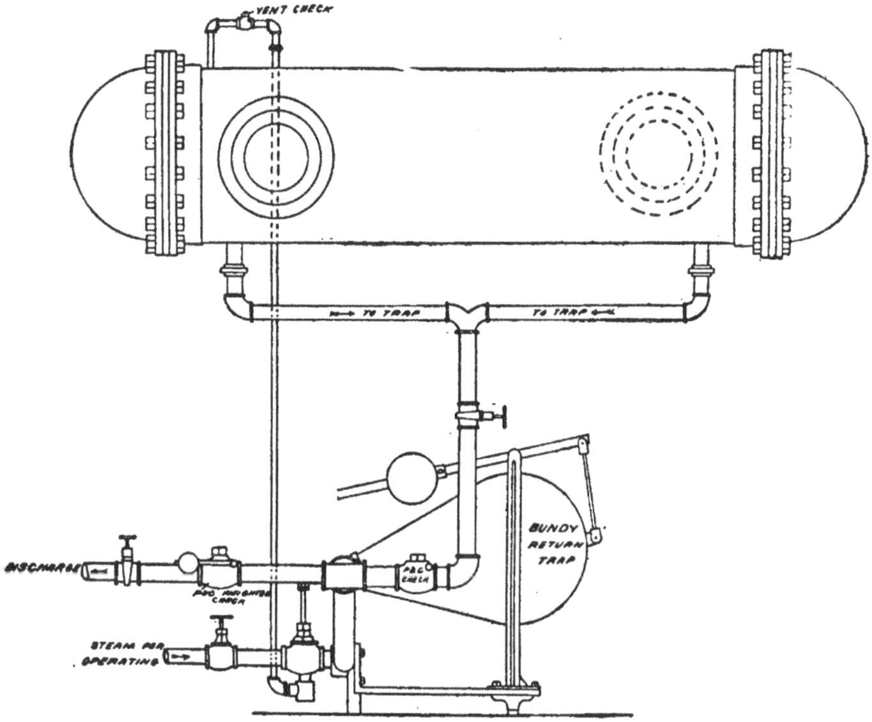

FIG. 65.—A Bundy Return Trap Connected to take Condensation from a Vacuum Chamber.

When returning water direct to a boiler, or range of boilers, the trap must be placed at least 4 ft. above the water level, as gravity is depended upon to effect the return to them from the traps. This, of course, is an essential condition for all return systems.

These traps will take condensation from coils situated either above or below the level upon which it may be rest ing, but it is conditional to successful operation that when the coils are below the trap there must be sufficient pressure available to lift the water into the trap, but when the coils, radiators, or whatever may be the systems under drainage, are located above the trap, the water will drain into it by gravity. One beauty of these return systems is that it is quite immaterial at what pressure above or below the boiler pressure may exist in the system under drainage, as they are capable of taking water from low pressure, or even gravity supply, and discharge the same into the boiler against a high pressure.

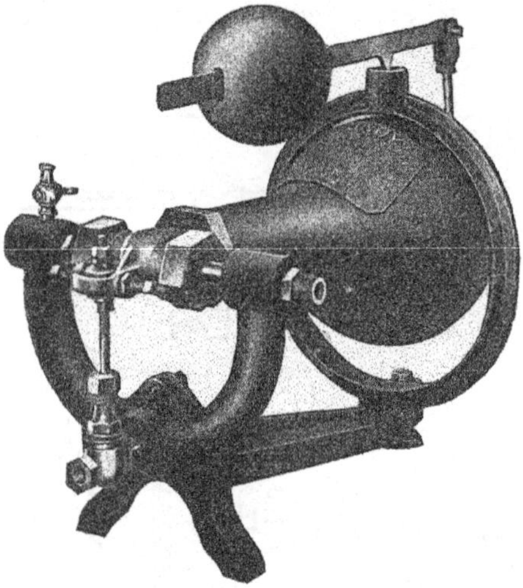

Fig. 66.—The Bundy 100 Series Separating Trap.

An interesting application of this trap is shown in fig. 65, where one is seen taking care of the discharge of both condensation and oil from a vacuum chamber, in which case it is equipped with a specially-designed set of valves suited to the conditions. The makers inform me that when properly set this trap will take water from a 25 in. or 26 in. vacuum without trouble or interference. A supply of live steam at nominal pressure—sufficient to evacuate the trap at the time of discharge—must be provided and connected as indicated.

The Bundy separating traps are very similar in appearance to the return traps, though their action is slightly different. An outside view of one is shown in fig. 66. They are mostly used for discharging condensation from coils or radiators either into the atmosphere or into a receiver. They are manufactured in two sizes, known by their makers as the 50 series and the 75 series respectively. The 50 series traps are designed to operate under all pressures up to 80 lb. per square inch, and the 75 series under higher-pressure steam, either saturated or superheated, up to 250 lb. per square inch. Their guiding principle is the differentiating counterbalance controlled by

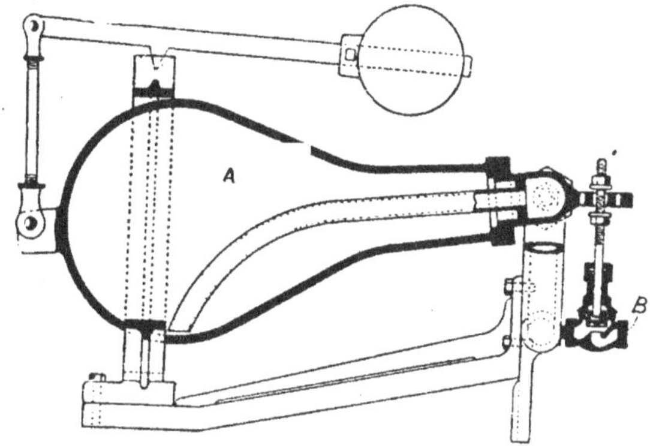

FIG. 67.—Sectional View—Bundy 100 Series Separating Trap.

the weight of water which collects in the receiving bowl, causing it to rise or fall, and in so doing operate a valve.

Fig. 67 shows the principle of construction, which is very similar to fig. 64, but the process of operation differs in some respects. The water enters through the trunnion, situated as in the return type, passes into the receiving bowl and fills it, when the latter falls and opens the discharge valve B. The pressure exerted upon the surface of the water, which is that of the system under drainage, forces the water out through the curved pipe A, *viâ* the yoke piece, to the discharge valve B. This discharge may be considered as purely a syphoning action. Before the bowl is quite empty, however, and while there is sufficient water remaining to seal the end of the pipe A, the balance

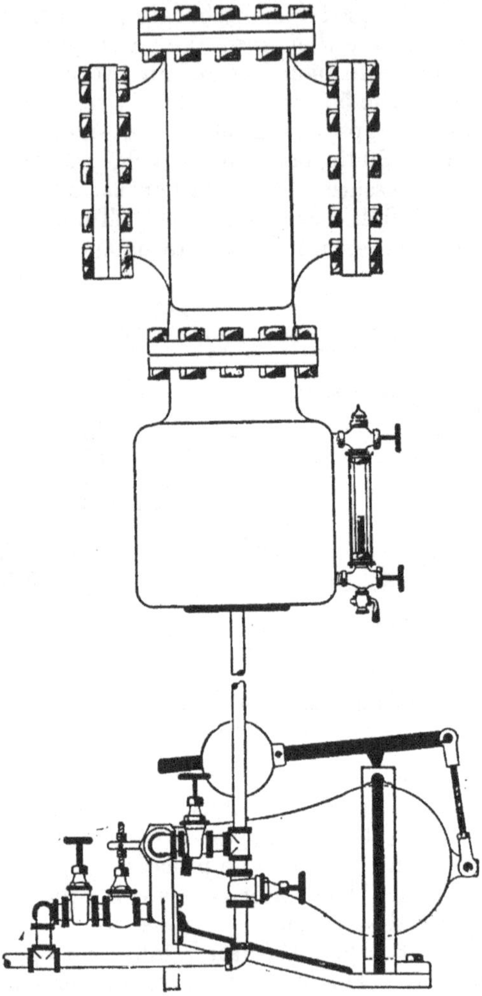

FIG. 68.—Bundy Separating Trap Draining a Separator.

weight causes the bowl to take up its original position and close the discharge valve.

Fig. 68 shows one of this type connected to a steam separator or drier. They should always be placed below the apparatus to be drained, and connected to a suitable by-pass substantially as indicated. They will raise water above their own level, in which case the usual non-return valves should be fitted. The discharge from these traps is intermittent and periodical, and will be in conformity with the length of time required for the receiving bowl to fill, this depending upon the nature of the system under drainage. No water can be discharged whilst the pear-shaped bowl is in the filling position, resting against the top of the hoop.

CHAPTER IX.

The Automatic Valves of the Atmospheric Steam Heating Company.

In order to render this treatise more complete, the writer has decided to include a few words by way of description of the special valves used upon the radiators, etc., of the system of atmospheric steam heating employed by the Atmospheric Steam Heating Company, of London, for these valves perform all the functions of a steam trap, albeit under somewhat different circumstances to those one generally associates with traps. It is interesting to mention, while on the subject, that the heating of large public buildings and works by the atmospheric system is very much on the increase, for the advantages in its favour, when compared with heating under pressure, are very considerable; among the most important may be mentioned the gain in efficiency and economy.

Fig. 69 shows a vertical cross-section through a valve used for extracting the water of condensation and air from a heating system, it being used in conjunction with a vacuum pump, the latter forming part of their system.

The position of the parts as shown in fig. 69 is that existing when the valve is closed, for the connection to the radiator, kettle, or other appliance under drainage, is made at C, and the outlet to the drain or tank at L through a vacuum pump or other draining apparatus. Briefly, its action is as follows: Water of condensation accumulating in the radiator, or other appliance, passes into the body E of the valve and rises, thereby sealing the space between the inverted cylinder-shaped shell A and

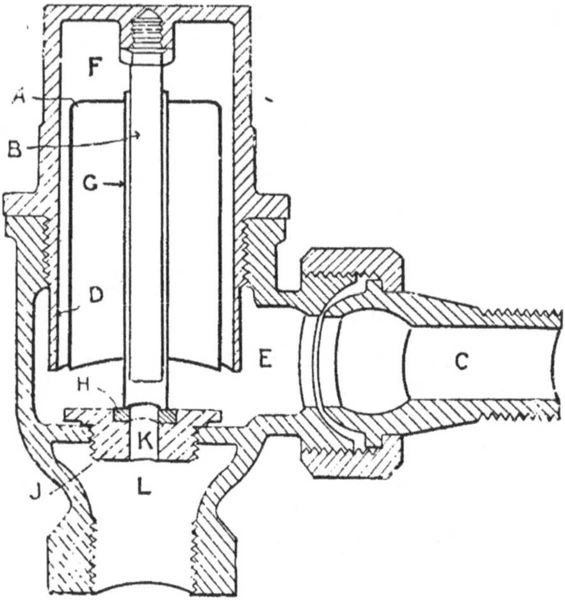

FIG. 69.—Water and Air Relief Valve of the Atmospheric Steam Heating Co.

the extension of the cover or bonnet D; when the difference of pressure acting on the shell A causes it to rise, lifting the lower end of the tube G of the vulcanised rubber disc H and permitting the accumulated condensation to pass away, the parts falling back again into place ready for another discharge. The vacuum pump or other draining agent extracts air from the system through the annular space between the central tube G and the guide spindle B, this space being always open to the system through the other annular space between D and A.

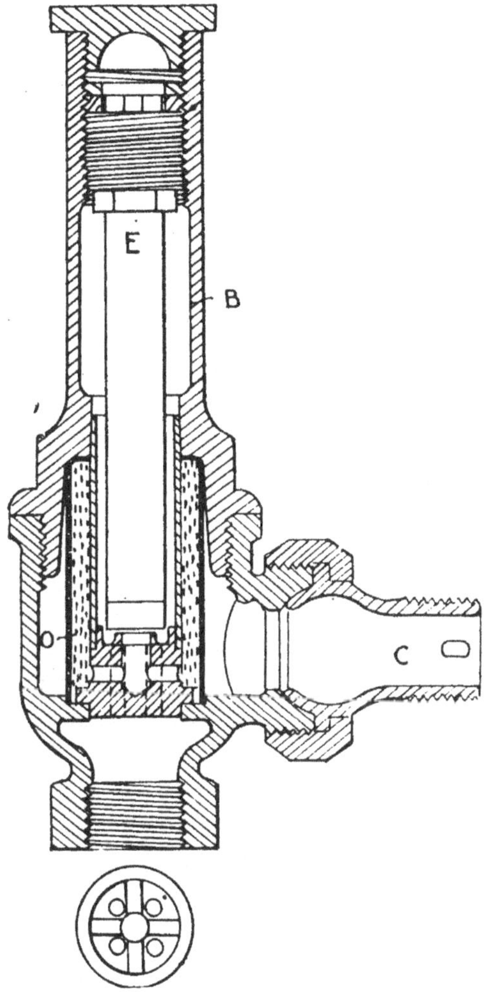

FIG. 70.—Thermostatic Relief Valve of the Atmospheric Steam Heating Co.

The reader will now see that the conditions of working are not so widely separated from those existing in steam trap practice as generally understood, for this is only a special type of steam trap applied to a special type of steam heating system.

Another valve, known as a Thermostatic Water and Air Relief Valve, which has also been specially designed for use in connection with this special heating system, is shown in fig. 70. It differs very considerably from the previous one, for it works upon an entirely different principle. As its name implies, it depends upon the thermal expansion and contraction of the centrally-situated composite plug E; for, being connected at C to the apparatus under drainage, the presence of water in the body B would cause this plug to contract sufficiently to allow the vacuum pump to suck it away through the specially-designed valve. The screen O prevents the passage of dirt to the holes in valve seating or bushing L. It is withdrawn with the bonnet B from the body for cleaning purposes. It should be noted that this type of water and air relief valve is only intended for use in positions where the circumstances entail a very small amount of condensation.

Yet another of these special valves is shown in fig. 71 This type is known as a "motor valve," and is specially intended for use under conditions entailing a somewhat excessive amount of condensation. Being of a more complex design, I propose to deal with it in greater detail, for its numerous cleverly-thought-out features cannot fail to be of more than passing interest.

It consists of the following parts: A substantially-constructed body, having an inlet at C and a discharge at L, and containing a hollow spherical copper float A, this latter being free to rise and fall on a tubular stem B. At the top end of this tubular stem, and firmly connected to it, is a movable piston F, whereas at the lower end is situated the discharge valve G; these latter also being contained in the body. The function of the copper float is to rise in the presence of water of condensation and close the port E in the piston F by means of the small valve D, this being necessary in order to relieve the upper side of the piston of any pressure in the system and to permit the vacuum caused by the vacuum pump to act on the upper side, this having the effect of raising the piston, together with the discharge valve, in order that a discharge may take place.

The action is as follows: Valve G being in the position shown, no direct communication to the outlet L can occur, but the vacuum induced in the pipe attached to L is

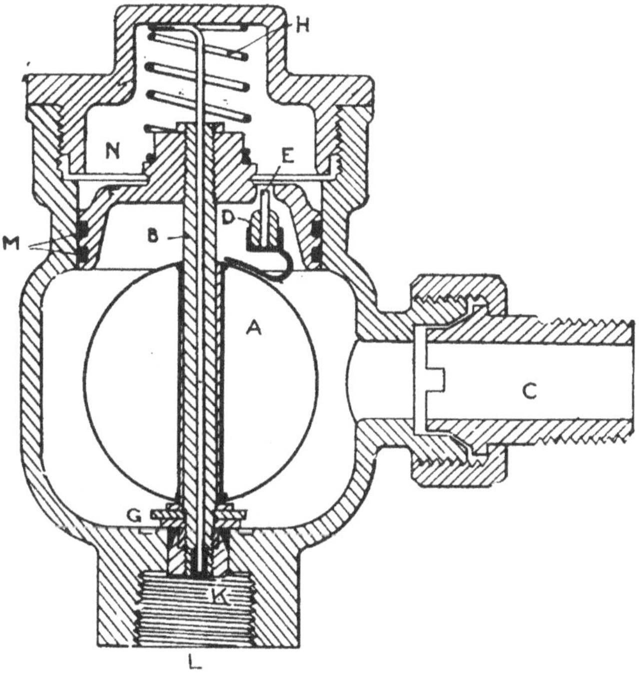

FIG. 71.—Motor Relief Valve of the Atmospheric Steam Heating Co.

communicated to the upper chamber N through the tube B. As the water of condensation entering at C accumulates, the float A rises and the valve D attached to it closes the port E. This prevents the vacuum existing in the chamber N from having any effect upon the lower chamber, and the unequal state of pressure set up between the two chambers, on either side of the piston, causes it to rise, lifting the valve G and permitting a discharge to take place. When the lower chamber is nearly emptied the float A drops back into its initial position and opens the port E, so that the pressure in the two chambers is again equalised, and the piston, aided by the spring H, falls back and closes the discharge valve. Air accumulating is continuously relieved through port E, chamber N, and tube B, except at such times when discharge is taking place.

INDEX.

John Heywood Ltd., Excelsior Printing and Bookbinding Works, Manchester.

Spotted something that isn't quite right?

Let us know and we'll fix it for future reprints. And as a thank you for your diligence, we'll send you a free Sothis book of your choice.

publisher@sothis.press

This page intentionally left blank

Made in the USA
Monee, IL
05 November 2026

67808938R00069